T0239422

SpringerBriefs in Energy

SpringerBriefs in Energy presents concise summaries of cutting-edge research and practical applications in all aspects of Energy. Featuring compact volumes of 50 to 125 pages, the series covers a range of content from professional to academic. Typical topics might include:

- A snapshot of a hot or emerging topic
- A contextual literature review
- A timely report of state-of-the art analytical techniques
- An in-depth case study
- A presentation of core concepts that students must understand in order to make independent contributions.

Briefs allow authors to present their ideas and readers to absorb them with minimal time investment.

Briefs will be published as part of Springer's eBook collection, with millions of users worldwide. In addition, Briefs will be available for individual print and electronic purchase. Briefs are characterized by fast, global electronic dissemination, standard publishing contracts, easy-to-use manuscript preparation and formatting guidelines, and expedited production schedules. We aim for publication 8–12 weeks after acceptance.

Both solicited and unsolicited manuscripts are considered for publication in this series. Briefs can also arise from the scale up of a planned chapter. Instead of simply contributing to an edited volume, the author gets an authored book with the space necessary to provide more data, fundamentals and background on the subject, methodology, future outlook, etc.

SpringerBriefs in Energy contains a distinct subseries focusing on Energy Analysis and edited by Charles Hall, State University of New York. Books for this subseries will emphasize quantitative accounting of energy use and availability, including the potential and limitations of new technologies in terms of energy returned on energy invested.

More information about this series at http://www.springer.com/series/8903

Samsul Ariffin Abdul Karim ·
Mohd Faris Abdullah · Ramani Kannan
Editors

Practical Examples of Energy Optimization Models

Springer

Editors
Samsul Ariffin Abdul Karim
Department of Fundamental
and Applied Sciences
Universiti Teknologi Petronas
Seri Iskandar, Perak, Malaysia

Mohd Faris Abdullah
Department of Electrical
and Electronic Engineering
Universiti Teknologi Petronas
Seri Iskandar, Perak, Malaysia

Ramani Kannan
Department of Electrical
and Electronic Engineering
Universiti Teknologi Petronas
Seri Iskandar, Perak, Malaysia

ISSN 2191-5520   ISSN 2191-5539   (electronic)
SpringerBriefs in Energy
ISBN 978-981-15-2198-0   ISBN 978-981-15-2199-7   (eBook)
https://doi.org/10.1007/978-981-15-2199-7

This Springer imprint is published by the registered company Springer Nature Singapore Pte Ltd.
The registered company address is: 152 Beach Road, #21-01/04 Gateway East, Singapore 189721, Singapore

Preface

Energy is the power which has the ability to perform any work. This energy is generally extracted from renewable and non-renewable sources. Sooner or later, the non-renewable energy maybe not available for us anymore. Therefore, scientist around the world are exploring the practicality of other sources that are called renewable sources of energy which is replenishable. In recent years, the growth of renewable energy integration with grid technology has been increased and breeding new questions for research and development (R&D). Inside the smart grid technology to integrate distributed energy sources efficiently to grid, several prediction control model and econometric model for predicting the global solar radiation and factors that affect solar radiation, performance evaluation of photovoltaic system, cost-benefit opportunity for end-use segment using lighting retrofit and improved energy consumption prediction model based on ANFIS approach have been undergoing for better sustainability in power generation from renewable energy sources around the world. In this book, several methods algorithm, environmental data-based performance analysis and experimental results and discussion will be covered to enhance broad readership regarding the topic. The reader from this book will know about the state of art about the pros and cons of renewable energy integration technology and suitability towards efficient power generation. The editors would like to express their gratitude to all the contributing authors for their great effort and full dedication in preparing the manuscripts for the book. Each chapter has been reviewed up to eight times by the reviewers and the editors. This is a very tedious process. We would like to thank all reviewers for reviewing all manuscripts and providing very constructive feedback.

This book is suitable for all postgraduate students and researchers who are working in this rapidly growing area. Any constructive feedback can be directed to the first editor.

Seri Iskandar, Malaysia

Samsul Ariffin Abdul Karim
Mohd Faris Abdullah
Ramani Kannan

Contents

Contributors

Elnazeer Ali Hamid Abdalla Electrical and Electronic Engineering Department, Universiti Teknologi PETRONAS, Seri Iskandar, Perak Darul Ridzuan, Malaysia

Mohd Faris Abdullah Electrical and Electronic Engineering Department, Universiti Teknologi PETRONAS, Seri Iskandar, Perak Darul Ridzuan, Malaysia

Chockalingam Aravind Vaithilingam Faculty of Innovation and Technology, School of Engineering, Taylor's University, Subang Jaya, Malaysia

Samuel Wong Weng Fai Electrical and Electronic Engineering Department, Universiti Teknologi PETRONAS, Seri Iskandar, Perak Darul Ridzuan, Malaysia

Reynato Andal Gamboa Lyceum of the Philippines University—Batangas, Capitol Site, Batangas, Philippines

Mohd Tahir Ismail School of Mathematical Sciences, Universiti Sains Malaysia, USM Minden, Penang, Malaysia

Ramani Kannan Electrical and Electronic Engineering Department, Universiti Teknologi PETRONAS, Seri Iskandar, Perak Darul Ridzuan, Malaysia

Samsul Ariffin Abdul Karim Fundamental and Applied Sciences Department and Centre for Smart Grid Energy Research (CSMER), Institute of Autonomous System, Universiti Teknologi PETRONAS, Seri Iskandar, Perak Darul Ridzuan, Malaysia

Perumal Nallagownden Electrical and Electronic Engineering Department, Universiti Teknologi PETRONAS, Seri Iskandar, Perak Darul Ridzuan, Malaysia

Nursyarizal Mohd Nor Electrical and Electronic Engineering Department, Universiti Teknologi PETRONAS, Seri Iskandar, Perak Darul Ridzuan, Malaysia

Mohd Fakhizan Romlie Electrical and Electronic Engineering Department, Universiti Teknologi PETRONAS, Seri Iskandar, Perak Darul Ridzuan, Malaysia

Then Yih Shyong Faculty of Innovation and Technology, School of Engineering, Taylor's University, Subang Jaya, Malaysia

Vaclav Skala Department of Computer Science and Engineering, Faculty of Applied Sciences, University of West Bohemia, Plzen, Czech Republic

Erman Azwan Yahya Electrical and Electronic Engineering Department, Universiti Teknologi PETRONAS, Seri Iskandar, Perak Darul Ridzuan, Malaysia

Muhammad Irfan Yasin Universiti Teknologi MARA (Terengganu) Kampus Kuala Terengganu, Kuala Terengganu, Terengganu Darul Iman, Malaysia

Fuzzy Regression Model to Predict Daily Global Solar Radiation

Muhammad Irfan Yasin, Samsul Ariffin Abdul Karim, Mohd Tahir Ismail and Vaclav Skala

The Fuzzy regression model provides a good alternative to the standard regression model that existing in statistics as well as engineering based studies. In this study, a new fuzzy regression model is introduced by incorporating the crisp and the spreading for the fuzziness of the data. The fuzzy triangular number is employed to obtain the fuzzy regression equation, i.e. left and right fuzzy quadratic regression model. This model will be used to predict the amount of solar radiation received at Universiti Teknologi PETRONAS (UTP), Malaysia. The performance of the proposed Fuzzy regression model is compared with standard polynomial quadratic fitting and Gaussian fitting. Based on the root mean square error (RMSE) and value of coefficient of determination, R^2, the proposed model is the best compared with those established regression models.

M. I. Yasin
Universiti Teknologi MARA (Terengganu) Kampus Kuala Terengganu, 21080 Kuala Terengganu, Terengganu Darul Iman, Malaysia
e-mail: muhammadnafri9601@gmail.com

S. A. A. Karim (✉)
Fundamental and Applied Sciences Department and Centre for Smart Grid Energy Research (CSMER), Institute of Autonomous System, Universiti Teknologi PETRONAS, Bandar Seri Iskandar, 32610 Seri Iskandar, Perak Darul Ridzuan, Malaysia
e-mail: samsul_ariffin@utp.edu.my

M. T. Ismail
School of Mathematical Sciences, Universiti Sains Malaysia, 11800 USM Minden, Penang, Malaysia
e-mail: m.tahir@usm.my

V. Skala
Department of Computer Science & Engineering, Faculty of Applied Sciences, University of West Bohemia, Univerzitni 8, CZ 306 14 Plzen, Czech Republic
e-mail: skala@kiv.zcu.cz

1 Introduction

Global solar radiation modelling is important in order to predict the amount of the solar radiation received at a certain location. This will help the person to install the solar panel on the building. Solar energy is an alternative to the conventional electricity. Solar radiation can be collected with a tool that located on top of the building. The most common tool for data acquisition is solarimeter [1–9]. To predict the amount of solar radiation, most researchers have used the quadratic polynomial fitting since the geometric shape of the solar data follows the parabolic form (Karim and Singh [4], and Khatib et al. [7]). Statistics property for the solar radiation time series is studied by Sulaiman et al. [8]. Jalil et al. [6] have studied the global solar radiation fitting with various fitting methods such as polynomial up to degree three, sine fitting, rational fitting as well as Gaussian fitting. They found that polynomial quadratic and Gaussian are the best method to predict the solar radiation. Gaussian fitting gives smaller root mean square error (RMSE) and higher coefficient of determination (R^2) compared with polynomial quadratic fitting. Others good references on data fitting methods can be found in Wang [10] and Hansen et al. [11]. So far in the literature, the model is developed based on polynomial quadratic fitting. There are many types of fuzzy number that can be used. For instance, triangular fuzzy number (TFN), Gaussian and trapezium fuzzy number (TrFN). Isa et al. [12] used the idea from Pan et al. [13, 14] i.e. using the multivariate matrix to find the best factor for fuzzy linear regression with symmetric triangular fuzzy numbers to select the best factor on the taxation in Malaysia. They use simple multivariate linear fitting by integrating the actual value as well as the spreading to find the width of the crisps. Xiao and Li [15] used the multi-dimensional fuzzy regression method to determine the influencing factors on the reality of the development of air cargo. It is proven that this method can effectively improve the accuracy of forecasting and reduce the risk of forecasting. Pan et al. [14] presents a fuzzy AHP approach to overcome the difficulties arising the suitable bridge construction from that other fuzzy AHP method involve complicated fuzzy mathematical calculations. Moreover, many of the existing fuzzy regression models require substantial computations due to complicated fuzzy arithmetic. Thus, fuzzy regression is a method that can be used to predict global solar radiation. This is the mainstream of the present study.

The main objective of the present studies is:

(a) To propose new quadratic fuzzy regression model
(b) To find the left and right fuzzy regression equations
(c) To predict the amount of solar radiation by using the proposed fuzzy regression model.

This chapter is organized as follows. In Sect. 1, some basic introduction as well as related literature review are presented. Section 2 is devoted to the construction of general fuzzy polynomial regression model by incorporating the crisp and spreading factor. Section 3, discusses the data collection and the tools that have been used to collect the data. Section 4 is dedicated for Results and Discussion including the comparison with established regression methods. In this section a new method to calculate the spreading value is proposed. This will overcome the weakness of the existing method to calculate the spreading for each solar radiation data. Section 5 explain the prediction model and Summary will be given in the final section.

2 Construction General Fuzzy Regression

The fuzzy regression for solar radiation data fitting involving several steps as shown in Fig. 1. It started with data collection using solarimeter until fuzzy regression model is developed and tested as well as the validation of the proposed method.

2.1 Fuzzy Regression

The general form of fuzzy regression model can be defined as

$$\tilde{y}_i = f(X, A) = \tilde{A}_0 + \tilde{A}_1 X_{i1}^2 + \cdots + \tilde{A}c_n X_{ik}^n \tag{1}$$

$\tilde{y}_i$ is the estimated fuzzy response variable or non-fuzzy output data. The function of $f(X, A)$ is consider which mapped from X into Y with the elements of X denoted by $x_i^{(t)} = \left(x_{i0}^t, x_{i1}^t, \ldots, x_{ik}^t\right)$ where $t = 1, 2, \ldots, n$ and ith which represents the independent or input variables of the model. The dependent variables or response

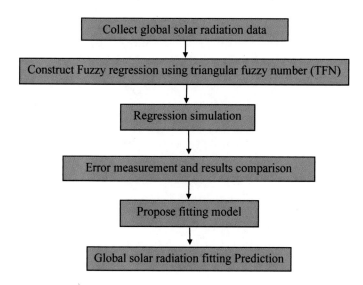

Fig. 1 Fuzzy regression modeling

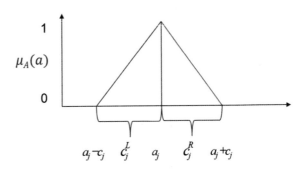

Fig. 2 Symmetrical Fuzzy regression coefficient

variables are denoted as Y_i in Y, where $\tilde{A} = (\tilde{A}_0, \tilde{A}_1, \ldots, \tilde{A}_n)$ are the model regression coefficients. If $\tilde{A}_j$ $(j = 0, 1, \ldots, n)$ are given as fuzzy sets, the model $f(X, A)$ is called a fuzzy model. $\tilde{A}_j$ is fuzzy coefficient in terms of symmetric fuzzy numbers.

The fuzzy coefficients of triangular fuzzy number (TFN) are shown in Fig. 2 which involve its centre or model value, left and right spreads. Asymmetrical TFN that represented by observed data is defined by a triplet $\tilde{y}_i = \left(a_j, c_j^L, c_j^R\right)$. The data was a symmetrical TFNs, thus $c_j^L = c_j^R = c_j$. Figure 2 also shows the membership function for the symmetrical fuzzy regression coefficients, $A_j = (a_j, c_j)$. The response variable for symmetric represented by $\tilde{y}_i = (a_j, c_j)$, where a_j is the centre point that represents the original value of data and c_j represents the spread.

The value of the spread, c_j is obtained from [8];

$$c_{n,j} = \frac{|y_i - \bar{y}|}{2} \qquad (2)$$

This method is considered a symmetrical triangular membership function in the following discussion. With k crisp independent variables and one fuzzy dependent variable, the estimated fuzzy quadratic regression can be expressed as

$$\tilde{y}_i = (a_0, c_{0,j}) + (a_1, c_{1,j})X_1 + (a_2, c_{2,j})X_2^2 + \cdots + (a_k, c_{k,j})X_k^n \qquad (3)$$

where $(a_0, c_{0,j})$ is the fuzzy intercept coefficient; $(a_1, c_{1,j})$ is the fuzzy slope coefficient for X_1; $(a_2, c_{2,j})$ is the fuzzy slope coefficient for X_2^2; $(a_k, c_{k,j})$ is the kth fuzzy slope coefficient.

The expected $\tilde{y}_i$ at a particular μ value is given by [15];

$$\mu_{\tilde{y},L} = \left[a_0 - (1 - \mu)c_{0,L}\right] + \left[a_1 - (1 - \mu)c_{1,L}\right]X_1 + \cdots$$
$$+ \left[a_k - (1 - \mu)c_{k,L}\right]X_k^n$$
$$= (a_0 + a_1 + \cdots + a_k) - (1 - \mu)\left(c_{0,L} + c_{1,L}X_1 + \cdots + c_{k,L}X_k^n\right) \qquad (4)$$

and

$$\mu_{\tilde{y},R} = (a_0 + a_1 + \cdots + a_k) + (1 - \mu)\left(c_{0,R} + c_{1,R}X_1 + \cdots + c_{k,R}X_k^n\right) \quad (5)$$

in which $a_0, a_1, \ldots, a_k$ are the estimated coefficients of $\tilde{y}_i$ at $\mu = 1$; $c_{0,L} + c_{1,L}X_1$ and $c_{0,R} + c_{1,R}X_1$ are the left fuzzy width and the right fuzzy width for X_1; $c_{0,L} + c_{2,L}X_2^2$ and $c_{0,R} + c_{2,R}X_2^2$ are the left fuzzy width and the right fuzzy width for X_2^2; $(1 - \mu)\left(c_{0,L} + c_{1,L}X_1\right)$ and $(1 - \mu)\left(c_{0,R} + c_{1,R}X_1\right)$ are the left fuzzy width and the right fuzzy width for X_1 at a given μ value; $(1 - \mu)\left(c_{0,L} + c_{2,L}X_2^2\right)$ and $(1 - \mu)\left(c_{0,R} + c_{2,R}X_2^2\right)$ are the left fuzzy width and the right fuzzy width for X_2^2 at a given μ value.

The general fuzzy quadratic model can be expressed in the following matrix form:

$$\tilde{y} = X\beta \quad (6)$$

where

$$\tilde{y} = \begin{bmatrix} \left(y_1, (1 - \mu)c_{1,j}\right) \\ \left(y_2, (1 - \mu)c_{2,j}\right) \\ \vdots \\ \left(y_n, (1 - \mu)c_{n,j}\right) \end{bmatrix} \quad (7)$$

$$X = \begin{bmatrix} 1 & x_{11} & x_{11}^2 \\ 1 & x_{12} & x_{12}^2 \\ \vdots & \vdots & \vdots \\ 1 & x_{1n} & x_{1n}^2 \end{bmatrix} \quad (8)$$

and

$$\beta = \begin{bmatrix} \beta_0 \beta_1 \vdots \beta_k \end{bmatrix} = \begin{bmatrix} \left(a_0, (1 - \mu)c_{0,j}\right) \\ \left(a_1, (1 - \mu)c_{1,j}\right) \\ \vdots \\ \left(a_n, (1 - \mu)c_{n,j}\right) \end{bmatrix} \quad (9)$$

In the above equations, matrices $\tilde{y}$ and X are the data matrices associated with response variable and predictor variables, respectively. Matrix β contains the least squares estimates of the regression coefficients as well as the width for each unknown.

To obtain the regression parameters, Eq. (6) can be rewritten as

$$\left(X'X\right)\beta = X'\tilde{y} \quad (10)$$

where X' is the transpose matrix of X.

The regression coefficients can be derived by matrix operations as follow:

$$\beta = \left(X'X\right)^{-1}X'\tilde{y} \quad (11)$$

where $\left(X'X\right)^{-1}$ is the inverse matrix of $X'X$ where,

$$X'X = \begin{pmatrix} n & \sum x_{1i} & \sum x_{1i}^2 \\ \sum x_{1i} & \sum x_{1i}^2 & \sum x_{1i}^3 \\ \sum x_{1i}^2 & \sum x_{1i}^3 & \sum x_{1i}^4 \end{pmatrix} \tag{12}$$

and

$$\tilde{y} = \begin{pmatrix} \sum y_i & \sum c_{i,j} \\ \sum x_{1i} y_i & \sum x_{1i} c_{i,j} \\ \sum x_{1i}^2 y_i & \sum x_{1i}^2 c_{i,j} \end{pmatrix}. \tag{13}$$

Besides that, Eq. (1) also can be solved by using Gaussian elimination or LU decomposition method.

2.2 Fuzzy Quadratic Regression Equation and Error Calculation

This section is adopted from Karim and Singh [4]. The error of deviation, e_i between actual value y_i and the estimates value $\hat{y}_i$ given as

$$e_i = y_i - \hat{y}_i$$

Three types of errors that can be calculated in fuzzy quadratic regression which are Sum Square Regression (SSR), Sum Square Errors (SSE) and Total of Sum Square (SST) where $\bar{y}_i$ stands for mean of fuzzy data, $\tilde{y}_i$. The equation of SSR, SSE and SST are shown in (15)–(17),

$$SSR = \sum_{i=1}^{n} \left(\tilde{y}_i - \bar{y}_i\right)^2 \tag{15}$$

$$SSE = \sum_{i=1}^{n} \left(y_i - \hat{y}_i\right)^2 = \sum_{i=1}^{n} e_i^2 \tag{16}$$

$$SST = SSR + SSE \tag{17}$$

The coefficient of determination R^2 is given by

$$R^2 = \frac{SSR}{SST} = 1 - \frac{SSE}{SST} \tag{18}$$

Table 1 ANOVA table for testing the significance of polynomial regression

Source of variation	Degree of freedom	Sum of squares	Mean square	Computed significance f
Regression	2	SSR	$\frac{SSR}{2}$	$f = \frac{SSR/2}{SSE/(n-3)}$
Residual	$n - 3$	$SSE = SST - SSR$	$\frac{SSE}{n-3}$	
Total	$n - 1$	SST		

R^2 measures the reduction in variability of y obtained using the regressors' $x_1, x_2, \ldots, x_k$ and $0 \leq R^2 \leq 1$. High value of R^2 suggest a good fitting to the actual data. Another statistical goodness fit used to measure the effectiveness of the proposed fuzzy quadratic fitting regression is RMSE defined as follows:

RMSE can be calculated by using the following formula,

$$RMSE = \sqrt{\frac{SSE}{n}} \tag{19}$$

whereas MSE is the error mean square given by SSE divided by its degrees of freedom, n. To test the hypothesis, the analysis of variance (ANOVA) table is used as shown in Table 1.

3 Data Collection

The global solar radiation data is collected at Universiti Teknologi PETRONAS (UTP) located at longitude 100.9382°E and latitude 4.3704°N at State of Perak, Malaysia. The population in UTP is approximately around 10,000 inclusive students and staffs. The data is collected for every 30 min from 0630 (6.30 am) till 1930 (7.30 pm) every day. Data were collected three times by using solarimeter and the data are gather by using computerized data acquisition system. In this study, the average of the global solar radiation data is used for one month. These actual data can be found in [6].

4 Results and Discussion

In this section, the capability of the proposed fuzzy quadratic regression to fit the global solar radiation data is investigated in detail. For the first fitting model based on fuzzy regression developed in the previous section, the spreading value, $c_{n,j}$ are calculated through Eq. (2). Table 2 shows the solar radiation data as well as its spreading value. X-coordinate indicate the time in hours. The spreading value, $c_{n,j}$

Table 2 Spreading value

X coordinate	Time (h)	True value, y_i	$c_{n,j}$
1	0630	0.0740	187.0246
2	0700	0.4500	186.8366
3	0730	42.7380	165.6926
4	0800	120.2580	126.9326
5	0830	216.6060	78.7586
6	0900	305.9940	34.0646
7	0930	413.6740	19.7754
8	1000	500.1440	63.0104
9	1030	577.9400	101.9084
10	1100	759.3760	192.6264
11	1130	847.9220	236.8994
12	1200	785.6440	205.7604
13	1230	712.3920	169.1344
14	1300	830.4120	228.1444
15	1330	687.4600	156.6684
16	1400	579.8000	102.8384
17	1430	605.7840	115.8304
18	1500	424.1780	25.0274
19	1530	439.7460	32.8114
20	1600	390.5980	8.2374
21	1630	337.9320	18.0956
22	1700	231.5060	71.3086
23	1730	155.5160	109.3036
24	1800	100.0700	137.0266
25	1830	34.4680	169.8276
26	1900	0.6440	186.7396
27	1930	0.0000	187.0616

varies and some are bigger than the actual solar radiation data. These may affect the accuracy of the proposed fitting fuzzy regression model later.

Now, by applying the proposed fuzzy polynomial quadratic fitting, the following matrix is obtained:

$$
\tilde{y} = \begin{bmatrix}
0.074 & 187.0246 \\
0.45 & 186.8366 \\
42.738 & 165.6926 \\
120.258 & 126.9326 \\
216.606 & 78.7586 \\
305.994 & 34.0646 \\
413.674 & 19.7754 \\
500.144 & 63.0104 \\
577.94 & 101.9084 \\
759.376 & 192.6264 \\
847.922 & 236.8994 \\
785.644 & 205.7604 \\
712.392 & 169.1344 \\
830.412 & 228.1444 \\
687.46 & 156.6684 \\
579.8 & 102.8384 \\
605.784 & 115.8304 \\
424.178 & 25.0274 \\
439.746 & 32.8114 \\
390.598 & 8.2374 \\
337.932 & 18.0956 \\
231.506 & 71.3086 \\
155.516 & 109.3036 \\
100.07 & 137.0266 \\
34.468 & 169.8276 \\
0.644 & 186.7396 \\
0 & 187.0616
\end{bmatrix}
\quad
X = \begin{bmatrix}
1 & 1 & 1 \\
1 & 2 & 4 \\
1 & 3 & 9 \\
1 & 4 & 16 \\
1 & 5 & 25 \\
1 & 6 & 36 \\
1 & 7 & 49 \\
1 & 8 & 64 \\
1 & 9 & 81 \\
1 & 10 & 100 \\
1 & 11 & 121 \\
1 & 12 & 144 \\
1 & 13 & 169 \\
1 & 14 & 196 \\
1 & 15 & 225 \\
1 & 16 & 256 \\
1 & 17 & 289 \\
1 & 18 & 324 \\
1 & 19 & 361 \\
1 & 20 & 400 \\
1 & 21 & 441 \\
1 & 22 & 484 \\
1 & 23 & 529 \\
1 & 24 & 576 \\
1 & 25 & 625 \\
1 & 26 & 676 \\
1 & 27 & 729
\end{bmatrix}
$$

From Eq. (11), the coefficients β can be calculated as below:

$$
\beta = \begin{pmatrix}
27 & 378 & 6930 \\
378 & 6930 & 142884 \\
6930 & 142884 & 3142062
\end{pmatrix}^{-1}
\begin{pmatrix}
10101.326 & 3317.3454 \\
135599.1 & 45223.64 \\
2046220.96 & 838054.361
\end{pmatrix}
$$

$$
\beta = \begin{pmatrix}
-231.71 & -0.8547 \\
132.08 & 2.8169 \\
-4.8441 & -0.1013
\end{pmatrix}
$$

The results above can be summarized in Tables 3, 4, 5 and 6.

Table 3 Fuzzy regression coefficient

	β_0	β_1	β_2
a_j	-231.71	132.08	-4.8441
$c_{n,j}$	-0.8547	2.8169	-0.1013

Table 4 Fuzzy quadratic regression form for membership function

μ	Fuzzy regression form
0	$y = (-231.71, -0.8547) + (132.08, 2.8169)x + (-4.8441, -0.1013)x^2$

Table 5 RMSE value and R^2 for fuzzy regression

μ	RMSE		R^2	
	y_L	y_R	y_L	y_R
0	283.7463	389.9918	0.0253	0.8412
1	105.9281		0.8746	

Table 6 Regression statistic

Multiple R	0.935207084
R square	0.87461229
Adjusted R square	0.864163314
Standard error	105.9281045
Observations	27

The fuzzy regression can be written as follows:

$$y_L = -400.4445 + 124.0027x - 5.1061x^2$$
$$y_R = -62.9687 + 140.1615x - 4.5821x^2$$

where y_L indicate the fuzzy regression for the left triangular number, meanwhile y_R is fuzzy right triangular number.

While, the crisp equation i.e. standard polynomial quadratic fitting is

$$y_{P2} = -231.71 + 13208x - 4.8441x^2.$$

Thus, instead of one fitting model obtained when using the standard quadratic polynomial, by using fuzzy regression, there are another two different quadratic fitting for the same data set. This is the advantage of fuzzy regression model for prediction and forecasting.

From Table 7 and Fig. 3, the resulting fuzzy regression is not superior compared with the standard quadratic fitting. The main reason for this is the choice of the spreading value from Eq. (2). To improve the performance of the fuzzy regression,

Table 7 Table of ANOVA for polynomial regression

Source of variation	Degree of freedom	Sum of squares	Mean square	Calculated f
Regression	2	1,878,426.675	939,213.3	83.70316
Residual	24	269,298.3197	11,220.76	
Total	26	2,147,724.995		

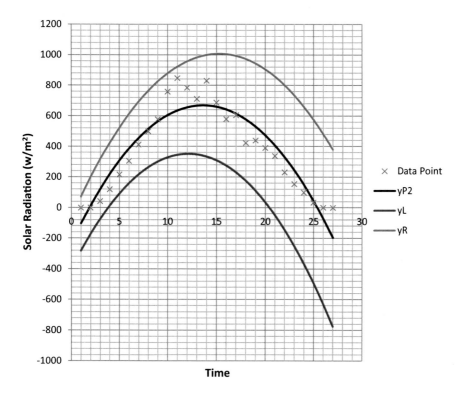

Fig. 3 Result for spreading value using given formula of $c_{n,j}$ with $\mu = 0$

different method is proposed to calculate the spreading value, $c_{n,j}$. The new spreading value is obtained by considering the error of the measurement of solarimeter. These new spreading values are shown in Table 8. Most of the spreading value is smaller than the actual solar radiation data. This may give a better fitting fuzzy regression model than the standard quadratic fitting model as well as the fuzzy regression fitting obtained from the previous section (Fig. 4; Tables 9, 10, 11).

Now, the coefficients β by using the new spreading value can be calculated as:

$$\beta = \begin{pmatrix} 27 & 378 & 6930 \\ 378 & 6930 & 142884 \\ 6930 & 142884 & 3142062 \end{pmatrix}^{-1} \begin{pmatrix} 10101.326 & 340 \\ 135599.1 & 4730 \\ 2046220.96 & 78410 \end{pmatrix}$$

Table 8 New spreading value

X coordinate	Time (h)	True value, y_i	New $c_{n,j}$
1	0630	0.0740	0
2	0700	0.4500	0
3	0730	42.7380	10
4	0800	120.2580	10
5	0830	216.6060	10
6	0900	305.9940	10
7	0930	413.6740	10
8	1000	500.1440	20
9	1030	577.9400	20
10	1100	759.3760	20
11	1130	847.9220	20
12	1200	785.6440	20
13	1230	712.3920	20
14	1300	830.4120	20
15	1330	687.4600	20
16	1400	579.8000	20
17	1430	605.7840	20
18	1500	424.1780	10
19	1530	439.7460	10
20	1600	390.5980	10
21	1630	337.9320	10
22	1700	231.5060	10
23	1730	155.5160	10
24	1800	100.0700	10
25	1830	34.4680	10
26	1900	0.6440	10
27	1930	0.0000	0

$$\beta = \begin{pmatrix} -231.71 & -0.8547 \\ 132.08 & 2.8169 \\ -4.8441 & -0.1013 \end{pmatrix}$$

Clearly with new spreading value, $c_{n,j}$, the fuzzy regression model is better than standard polynomial quadratic fitting (crisp). Furthermore, Fuzzy regression with membership value $\mu = 0.7$ give better results (both for left and right fuzzy triangular numbers) than the other membership value. When the data is highly fuzzy i.e. at $\mu = 0$, the left and right fuzzy regression also better than standard polynomial quadratic fitting. The result from the proposed fuzzy regression model can also be further improved by utilizing both fuzzy regression i.e. left and right regression. To do

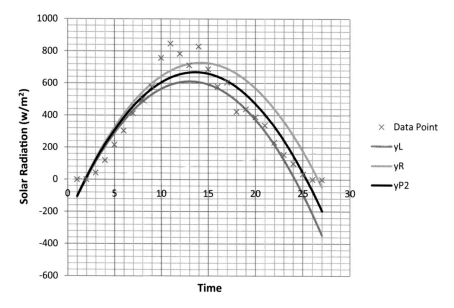

Fig. 4 Fitting curves for spreading value using new $c_{n,j}$ with $\mu = 0$

Table 9 Table of ANOVA for $\mu = 0$

Source of variation	Degree of freedom	Sum of squares		Mean square		Calculated f	
		y_L	y_R	y_L	y_R	y_L	y_R
Regression	2	1,900,324.67	1,718,884.46	950,162.33	859,442.23	92.17	48.09
Error	24	247,400.32	428,840.53	10,308.34	17,868.35		
Total	26	2,147,724.995					

Table 10 RMSE value and R^2 for fuzzy regression

μ	y_L		y_R	
	RMSE	R^2	RMSE	R^2
0	97.5469	0.8848	128.4284	0.8003
0.1	94.9254	0.8909	123.1407	0.8164
0.2	92.6836	0.8960	118.1521	0.8310
0.3	90.8550	0.9001	113.4495	0.8442
0.4	89.4641	0.9031	109.0680	0.8560
0.5	88.5307	0.9051	105.0320	0.8665
0.6	88.0740	0.9061	101.3727	0.8756
0.7	88.0640	0.9061	98.0813	0.8835
0.8	88.4204	0.9054	95.1959	0.8903
0.9	89.3360	0.9034	92.7442	0.8959

Data fitting method	Statistical goodness fit	
Degree	RMSE	R^2
2	105.93	0.8746

Table 11 RMSE value and R^2 for polynomial fitting

this, the convex combination between y_L and y_R fuzzy regression is employed. This formulation is given as follows:

$$\tilde{y}_a = (1 - a)y_L + ay_R \tag{20}$$

with y_L is the fuzzy equation for left spread and y_R is the fuzzy equation for right spreading with the same membership number and a is a free parameter and positive.

The fuzzy convex combination form for membership function, $\mu = 0$ and $a = 0.3$ is given as

$$\tilde{y}_{a=0.3} = 0.7\left(-232.5613 + 129.2652x - 4.9454x^2\right)$$
$$+ 0.3\left(-230.8519 + 134.899x - 4.7428x^2\right)$$

Figure 5 shows the examples of fitting curve using convex combination method with $a = 0.3$ and $\mu = 0$ (Tables 12 and 13).

The prediction of fuzzy convex combination form for membership function, $\mu = 0$ when $a = -1.5$ and $a = 1.5$ is given in Table 14.

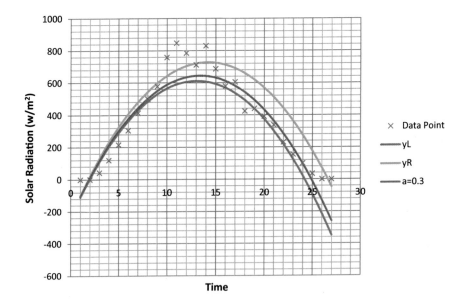

Fig. 5 Fitting model using convex method with $a = 0.3$

Table 12 ANOVA table for $a = 0.3$

Source of variation	Degree of freedom	Sum of squares	Mean square	Calculated f
Regression	2	1,946,471.166	973,235.5832	116.06
Error	24	201,253.8285	8385.576188	
Total	26	2,147,724.995		

Table 13 RMSE value and R^2 for fuzzy convex combination

a	$(1 - a)y_L + ay_R$	
	RMSE	R^2
0.1	90.76613163	0.89643
0.2	87.57157559	0.903592
0.3	86.33565602	0.906294
0.4	87.14175005	0.904536
0.5	89.93496625	0.898318
0.6	94.53933657	0.88764
0.7	100.7067456	0.872502
0.8	108.1701693	0.852904
0.9	116.6811762	0.828846

Table 14 RMSE value and R^2 for fuzzy convex combination

a	$(1 - a)y_L + ay_R$	
	RMSE	R^2
-1.5	216.4865	0.410822
1.5	180.408	0.590837

When the value of the free parameter is -1.5 (negative) as well as 1.5 (greater than 1), the fitting is not good, and this is the main reason why the best fitting model can be obtained when a is between 0 and 1. Figure 6 shows the combination of all fitting techniques including the crisp denoted as yP2. Clearly from the graphs, the proposed fitting model with $a = 0.3$, give the best results.

5 Prediction Global Solar Radiation

Since between 1100 until 1600, the solar radiation received is very high, the weighted fuzzy regression model constructed in Sect. 4 is used to predict the solar radiation received at UTP. Firstly, when membership value $\mu = 0.7$ the prediction is given in Table 15.

Next the prediction of the solar radiation by using the proposed convex (weighted) combination with $a = 0.3$ when $\mu = 0$ is shown in Table 16.

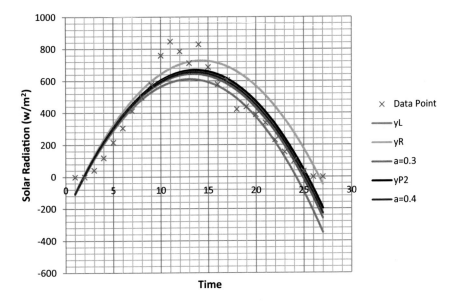

Fig. 6 Fitting curves for the best spreading value with $\mu = 0$

Table 15 Fuzzy quadratic regression form $\mu = 0.7$

X coordinate	Time (h)	y_{P2}	y_L	y_R
10.5	1115	621.067975	608.586425	633.549525
11.5	1145	646.577775	632.582325	660.573225
12.5	1215	662.399375	646.829225	677.969525
13.5	1245	668.532775	651.327125	685.738425
14.5	1315	664.977975	646.076025	683.879925
15.5	1345	651.734975	631.075925	672.394025
16.5	1415	628.803775	606.326825	651.280725
17.5	1445	596.184375	571.828725	620.540025
18.5	1515	553.876775	527.581625	580.171925
19.5	1545	501.880975	473.585525	530.176425

6 Summary

This study considers the fuzzy regression model to predict the global solar radiation data. The fuzzy triangular number is used in order to construct the fuzzy regression for left, right and crisp (centre). To apply the fuzzy regression, the spreading value $c_{n,j}$ is calculated based on formula given in [8]. Based on the numerical results, the spreading value calculated by using Eq. (2) does not give better fuzzy regression model since the error quite high. To improve the results, a new spreading value is

Table 16 Fuzzy quadratic regression with $a = 0.3$ for $\mu = 0$

X coordinate	Time (h)	y_L	y_a	y_R
10.5	1115	579.49295	604.453235	662.6939
11.5	1145	599.95935	627.946935	693.2513
12.5	1215	610.53495	641.671395	714.3231
13.5	1245	611.21975	645.626615	725.9093
14.5	1315	602.01375	639.812595	728.0099
15.5	1345	582.91695	624.229335	720.6249
16.5	1415	553.92935	598.876835	703.7543
17.5	1445	515.05095	563.755095	677.3981
18.5	1515	466.28175	518.864115	641.5563
19.5	1545	407.62175	464.203895	596.2289

proposed. Numerical simulation using the new spreading value gave a better model compared with standard polynomial quadratic fitting i.e. crisp model. The left fuzzy regression is always better than the right fuzzy regression. But to accommodate both left and right regression models, the convex (weighted) combination method is employed. Based on the results, the fuzzy regression with $a = 0.3$ give the best solar radiation fitting and is used to predict the amount of solar radiation for every 15 min between 1100 until 1600 h. Future work will be focusing on the development of fuzzy multivariate regression for more than one independent variables which is our main objective for future.

Acknowledgements This study is fully supported by Universitas Islam Riau (UIR), Pekanbaru, Indonesia and Universiti Teknologi PETRONAS (UTP), Malaysia through **International Collaborative Research Funding (ICRF): 015ME0-037**. The first author is currently doing his internship at UTP under Research Attachment Program (RAP).

References

1. Karim SAA, Singh BSM, Razali R, Yahya N (2011) Data compression technique for modeling of global solar radiation. In: Proceeding of 2011 IEEE international conference on control system, computing and engineering (ICCSCE) 25–27 Nov 2011, Holiday Inn, Penang, pp 448–352
2. Karim SAA, Singh BSM, Razali R, Yahya N, Karim BA (2011) Solar radiation data analysis by using Daubechies wavelets. In: Proceeding of 2011 IEEE international conference on control system, computing and engineering (ICCSCE) 25–27 Nov 2011, Holiday Inn, Penang, pp 571–574
3. Karim SAA, Singh BSM, Razali R, Yahya N, Karim BA (2011) Compression solar radiation data using Haar and Daubechies wavelets. In: Proceeding of regional symposium on engineering and technology 2011, Kuching, Sarawak, Malaysia, 21–23 Nov 2011, pp 168–174
4. Karim SAA, Singh BSM (2013) Global solar radiation modeling using polynomial fitting. Appl Math Sci 8:367–378

5. Karim SAA, Singh BSM, Karim BA, Hasan MK, Sulaiman J, Janier Josefina B, Ismail MT (2012) Denoising solar radiation data using Meyer wavelets. AIP Conf Proc 1482:685–690. https://doi.org/10.1063/1.4757559
6. Jalil MAA, Karim SAA, Baharuddin Z, Abdullah MF, Othman M (2018) Forecasting solar radiation data using Gaussian and polynomial fitting methods. In: Sulaiman SA, Kannan R, Karim SAA, Nor NM (eds) Sustainable electrical power resources through energy optimization and future engineering. Springer Briefs in Energy. Springer Nature Singapore Pte. Ltd.
7. Khatib T, Mohamed A, Sopian K (2012) A review of solar energy modeling techniques. Renew Sustain Energy Rev 16:2864–2869
8. Sulaiman MY, Hlaing Oo WM, Wahab AM, Sulaiman MZ (1997) Analysis of residuals in daily solar radiation time series. Renew Energy 29:1147–1160
9. Wu J, Chan CK (2011) Prediction of hourly solar radiation using a novel hybrid model of ARMA and TDNN. Sol Energy 85:808–817
10. Wang Y (2012) Statistics & applied probability. In: Smoothing splines: methods and applications. Chapman and Hall/CRC
11. Hansen PC, Pereyra V, Scherer G (2012) Least squares data fitting with applications. The Johns Hopkins University Press
12. Isa NHM, Othman M, Karim SAA (2018) Multivariate matrix for fuzzy linear regression model to analyze the taxation in Malaysia. Int J Eng Technol 7(4.33):78–82
13. Pan NF (2008) Fuzzy AHP approach for selecting the suitable bridge construction method. Autom Constr 17:958–965
14. Pan NF, Lin TC, Pan NH (2009) Estimating bridge performance based on a matrix-driven fuzzy linear regression model. Autom Constr 18:578–586
15. Xiao M, Li C (2018) Fuzzy regression prediction and application based on multi-dimensional factors of freight volume. IOP Conf Ser Earth Environ Sci 108:032071. https://doi.org/10.1088/1755-1315/108/3/032071

Performance Evaluation of a 2-kWp Grid Connected Solar Photovoltaic System at Gate 3 Universiti Teknologi PETRONAS

Mohd Faris Abdullah, Mohd Fakhizan Romlie, Samsul Ariffin Abdul Karim and Samuel Wong Weng Fai

As of now, energy demand of the world increases year by year and many nations are facing with the critical environmental issues related to conventional energy sources. Along this line, utilities and clients have broadly acknowledged the utilization of contamination-free sustainable energy-source based power generations, including solar energy. Solar energy is an alluring choice and has attracted a great attention and consideration since the most recent couple of decades because of the significant price drop in photovoltaic (PV) cells. In recent year, the installation of solar PV system is increasing annually in Malaysia as solar energy is generally acknowledged as one of the high potential renewable energy that can be implemented widely in this country. However, due to the uncertainty of solar radiation and weather pattern, the solar power is considered as an unsteady energy source. Consequently, performance forecast of this system is very important in numerous related aspects, for example, estimating and control of the system. Furthermore, solar PV system performance is important for system planning and financing as well as energy market analysis, particularly when this system is connected with the national power system grid. The main objective of this study is to evaluate the performance of the 2-kWp grid connected solar PV system, which is located at Universiti Teknologi PETRONAS

M. F. Abdullah (✉) · M. F. Romlie · S. W. W. Fai
Electrical and Electronic Engineering Department, Universiti Teknologi PETRONAS,
Bandar Seri Iskandar, 32610 Seri Iskandar, Perak Darul Ridzuan, Malaysia
e-mail: mfaris_abdullah@utp.edu.my

M. F. Romlie
e-mail: fakhizan.romlie@utp.edu.my

S. W. W. Fai
e-mail: wengfai94@hotmail.com

S. A. A. Karim
Fundamental and Applied Sciences Department and Centre for Smart Grid Energy Research
(CSMER), Institute of Autonomous System, Universiti Teknologi PETRONAS,
Bandar Seri Iskandar, 32610 Seri Iskandar, Perak Darul Ridzuan, Malaysia
e-mail: samsul_ariffin@utp.edu.my

© The Author(s), under exclusive license to Springer Nature Singapore Pte Ltd. 2020   19
S. A. A. Karim et al. (eds.), *Practical Examples of Energy Optimization Models*,
SpringerBriefs in Energy, https://doi.org/10.1007/978-981-15-2199-7_2

(UTP) and commissioned in June 2016. The relationship between its performance with external factor such as solar irradiance and PV module temperature will also be studied. The result of this study is significant and important when planning for large scale solar power generation in the future.

1 Introduction

In the developing stage of a country, energy plays a very important role. Lack of electricity and incomplete power supply in certain area especially rural area of a country may intensify poverty problem of the country. Malaysia is located at equatorial region, with 30 °C of average ambient temperature and about 4500 Wh/m^2 of average solar irradiation per day. Southwest and northeast monsoons that usually occur during April to October and October to February every year, respectively have bearing on the weather pattern. By having this geographical advantage, there are various of sustainable energy resources available in Malaysia. However, most of these sources are not fully capitalized due to lack of technology. More research are needed to improve the efficiency of these renewable energy sources [1].

In [1] also, the author studied the possibility and feasibility of rural electrification in Malaysia by applying sustainable energy sources, such as solar, wind and hydro. The author concluded that the renewable energies with highest possibility to be implemented in Malaysia are hydropower and solar power. However, there are some obstacles on the implementation of renewable energy in Malaysia, which can be classified into three divisions:

(1) Economic
(2) Law and legislation
(3) Monetary and organization.

Energy is a very important factor in the development and financial system of a country, and therefore a lot of research are taking place to satisfy the energy need of the country especially in the rural area. In [2], authors discussed about the energy demand in islanded rural areas, the applications of hybrid electricity generation systems, especially in renewable energy aspects, pros and cons of the renewable energy, environmental issues, and economic constitutions in a country. Renewable energy is a very good way to generate electricity for rural areas. Application of hybrid of the renewable energy will be more effective.

2 Solar Energy

Solar energy is energy created by the heat and light of the sun. Solar power is produced when this energy is converted into electric power or utilized to heat air, water, or other substances. There are currently two main types of solar energy technology namely solar thermal and solar PV.

2.1 Solar Thermal

Solar thermal technology has been utilized by human being since long time ago mainly for domestic usage such as space warming or to heat water. Today, solar thermal utilization has advanced to some industrial activities. Concentrated solar plant (CSP) depends on four fundamental frameworks which are collector, receiver (absorber), deliver/stockpiling and power conversion. The four primary technology that are commercially accessible today are:

(1) Parabolic trough
(2) Solar towers
(3) Parabolic dishes
(4) Linear Fresnel reflectors.

2.2 Solar PV

A PV cell, generally called a solar cell, is a non-mechanical equipment which is installed in a solar panel that converts sunlight directly into electric power. Solar PV technology has very high potential especially in islanded or rural areas where connection to national grid is very expensive. However, the cost of power generation through solar PV technology is too high, which is about 20 times higher than the cost of conventional power generations [3].

3 Solar Energy in Malaysia

Solar energy is the renewable energy source which has been recognized with the highest possibility to fulfill the energy demand as Malaysia located in the equatorial region, receives large amount of solar irradiance per year. The meaning of grid connected is that interconnection is assembled between the solar PV equipment with the national grid. However, high cost of solar PV equipment is the main obstruction or setback to apply this technology in Malaysia.

To study the possibility of application of this renewable energy, government of Malaysia (GoM) has initiated the Malaysia Building Integrated Photovoltaic (MBIPV) Technology Application Project in year 2005. In rural areas, especially in certain areas of East Malaysia, GoM has initiated the Rural Electrification Program (REP) to electrify rural areas using renewable energy.

Under Fit-in Tariff (FiT) scheme, a fixed tariff per unit of electricity exported into the national grid will be paid to the renewable energy producers. This is one of the way to stimulate the growth of renewable energy in Malaysia [4].

To keep up with the pace of the world, there are some financial aid facilities provided by the GoM to stimulate the development of sustainable energy technology in Malaysia. For instance, Green Technology Financing Schemes (GTFS) was launched during year 2010 to promote the applications of renewable energy among the companies and commercial sectors [4].

4 Solar PV Performance Indicators

The IEC 61724 standard is usually taken as reference to assess the performance of solar PV system. The parameters that will be evaluated are explained in the following sections [5].

4.1 Energy Output

The total energy is defined as the amount of alternating current (AC) power generated by the system over a given period of time. The total energy produced can be determined respectively in term of hourly, daily and monthly as [5]:

$$E_{AC,h} = \sum_{t=1}^{60} E_{AC,t} \tag{1}$$

$$E_{AC,d} = \sum_{h=1}^{24} E_{AC,h} \tag{2}$$

$$E_{AC,m} = \sum_{d=1}^{N} E_{AC,d} \tag{3}$$

where

$E_{AC,t}$ AC energy output at time t (in min);
$E_{AC,h}$ AC energy output at hour;
$E_{AC,d}$ daily AC energy output;
$E_{AC,m}$ monthly AC energy output;
N   the number of days in a month.

4.2 Array Yield

The array yield, Y_A is defined as the total DC (direct current) energy produced by the solar PV array over a designated time range normalized by the solar PV rated power ($P_{PV,rated} = 2$ kWp) [6]. The real output of the solar PV array is indicated by the array yield [7]. It is given as:

$$Y_A = \frac{E_{DC}}{P_{PV,rated}} (kWh/kW_P) \tag{4}$$

where

E_{DC} DC energy produced (kWh) by the solar PV array.

4.3 Final Yield

The final yield, Y_F is defined as the total AC energy produced by the solar PV system for a designated time range with respect to the rated output power [5]. It is given as:

$$Y_F = \frac{E_{AC}}{P_{PV,rated}} \left(\frac{kWh}{kW_P} \right) \tag{5}$$

where

E_{AC} AC energy output (kWh) of the solar PV system.

4.4 Reference Yield

The reference yield, Y_R is the ratio of the total in-plane solar irradiation, H_T to PV module's reference in-plane solar irradiance ($G_0 = 1$ W/m^2). It is given as [8]:

$$Y_R = \frac{H_T}{G_0} \left(\frac{\frac{Wh}{m^2}}{\frac{W}{m^2}} \right) \tag{6}$$

4.5 Solar PV Module Efficiency

The solar PV module array efficiency, n_{PV} represents the average energy conversion efficiency of the solar PV array, which is the ratio of daily array energy output (DC),

E_{DC} to the product of total daily in-plane irradiation, H_T and area of the solar PV array, A_m [9]. The solar PV module efficiencies, n_{PV} is given as:

$$(\%)n_{PV} = \frac{E_{DC} \times 100}{H_T \times A_m} \qquad (7)$$

where

E_{DC} daily array DC energy output
H_T total daily in-plane irradiation
A_m area of the solar PV array.

4.6   Inverter Efficiency

The inverter efficiencies, n_{INV} is given as:

$$(\%)n_{INV} = \frac{E_{AC} \times 100}{E_{DC}} \qquad (8)$$

where

E_{AC} daily array AC energy output.

4.7   Performance Ratio

The performance ratio is the ratio of the actual generation data of the solar PV system to the total solar irradiance received [10]. Performance ratio reflects the actual system efficiency and difference as compared to the nominal performance [5]. The calculation of performance ratio is unaffected by the location, so it is the best indicator to compare the performance of same or different solar PV module technology although the power rating is different [11]. Performance ratio is given as:

$$(\%)\,PR = \frac{Y_F \times 100}{Y_R} \qquad (9)$$

4.8   Capacity Factor

The capacity factor is defined as the energy delivered by an electric power generating system at full rated power, $P_{PV,rated}$ over a given period of time [12]. Unity can be obtained if the system produces full rated power non-stop [7]. The capacity factor of the solar PV system is given as:

$$(\%) \, CF_{annual} = \frac{E_{AC} \times 100}{P_{PV,rated} \times 24 \times 365} \tag{10}$$

$$(\%) \, CF_{monthly} = \frac{E_{AC} \times 100}{P_{PV,rated} \times 24 \times 30} \tag{11}$$

5 Data Collection

The 2-kWp grid connected solar PV system at Gate 3 UTP (latitude 4° 22′ 23.12″N, longitude 100° 58′ 10.39″E and 5 m above sea level) is designed as a ground mounting PV system as shown in Fig. 1. The solar panels are mounted on aluminum rail and rise-up by galvanized structure. This type of PV installation is commonly used in solar farm and flat roof building which required structure to rise-up the PV panel from the ground.

DC cables connect solar panels to DC/AC junction box where it plays role of enclosing DC and AC protecting devices. Devices in DC/AC junction box are connected to the inverter by DC cables. Inverter plays significant role in a PV system that converts DC power into AC power, which we are using every day. Inverter is working in such a way that it matches frequency and voltage with grid specifications, so that there will be no damage to power users. Finally, the inverter is connected to the

Fig. 1 The 2-kWp grid connected solar PV system at Gate 3 UTP

Fig. 2 The inverter and DC/AC junction box

main distribution box (DB) of the building. Figure 2 shows the picture of the inverter and DC/AC junction box. The technical specifications of the solar PV module and inverter are shown in Table 1.

The period of this study is chosen to be from September 2017 to August 2018 for one year of data collected from the data logger of the 2-kWp grid connected solar PV system at Gate 3 UTP. The collected data are solar PV module DC energy output, inverter AC energy output, solar irradiance and PV module temperature. All data are logged in the data logger and the sampling interval is 5 min.

6 Data Analysis

AC and DC energy outputs can be obtained from the logged data directly. The array yield, final yield and reference yield are calculated by using Eqs. (4), (5), and (6) respectively. Equations (7) and (8) are used to obtain the solar PV module and inverter efficiencies. Lastly, the performance ratio and capacity factor of the grid connected solar PV system are determined using Eqs. (9) and (10) respectively.

7 Solar PV Performance

Figure 3 shows the monthly DC and AC energy outputs and the solar irradiance received by the grid connected solar PV system. The lowest monthly DC and AC

Table 1 Technical specifications of solar PV module and inverter

1. Modules specifications	
Manufacturer	Sharp
Model	ND-R250A5
Technology	Polycrystalline (mc-Si)
STC power (Wp)	250
Vpmax (V)	30.9
Isc (A)	8.1
STC efficiency (%)	15.2
Modus count	8
Number of strings	1
Solar panels surface (m^2)	13.14
Weight per m^2 (kg/m^2)	11.57
Total panels weight (kg)	152
2. Inverter	
Manufacturer	ABB
Model	UNO-2.0-I-OUTD
Nominal input power (W)	2000
Max input voltage (V)	520
MPPT voltage range, full power (V)	200–470
Operating voltage range	140–520
Max. input current per MPPT (A)	15
Max. DC input power per MPPT(W)	2300
Nominal output power (VA)	2000
Maximum output current (A)	10.5
AC voltage range (V)	180–264
Maximum efficiency	96.3%
Cooling	Natural convection
IP degree of protection	IP65 (electronics and balance)

energy outputs were 183.6 kWh and 169.3 kWh respectively that occurred in January 2018. Meanwhile, the highest monthly DC and AC energy outputs were 286.3 kWh and 268.6 kWh respectively that experienced in March 2018. In January 2018, the amount of solar irradiation received by the solar PV was 99.9 kWh/m^2 while 159.4 kWh/m^2 of solar irradiation harvested in March 2018.

Obviously, there is a direct correlation between solar irradiation received by the solar PV and monthly DC and AC energy outputs. The irradiation received by the solar PV is weather dependent since more rainy days recorded in January 2018 as

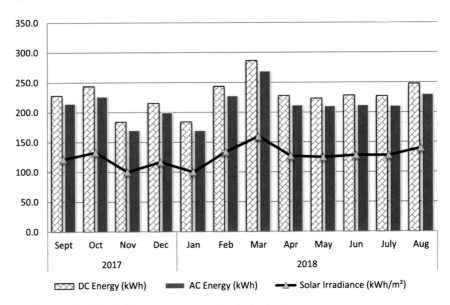

Fig. 3 Monthly DC and AC energy outputs and solar irradiation

compared to other months. On the other hand, March 2018 experienced the most sun light days as compared to the other months. The solar irradiation trending depicts the DC and AC energy outputs.

It can be noticed that the AC energy output is less than DC energy output and this difference constitute the energy loss during conversion from DC to AC energy. This energy loss can be analyzed by calculating the solar PV module and inverter efficiencies. The monthly solar PV module and inverter efficiencies are almost constant as shown in Fig. 4. The monthly average efficiency of the solar PV module was 13.8%. Solar PV supplier stated that the efficiency of their PV cell is 15.2% at standard test condition (STC). The efficiency of a PV module is always less than a cell since the amount of energy hitting the individual cell is not the same as the whole PV module.

The other factors that can cause solar PV module less efficient are dust and dirt accumulated on the solar panels which limit the solar irradiance absorption, weather condition, panel orientation, etc. Furthermore, the solar PV module is not operating at STC which can further reduce its efficiency. The average efficiency of the inverter was 93.1% while the maximum efficiency declared by the manufacturer is 96.3%. Among others, the inverter efficiency depends on the DC voltage, irradiance level, inverter temperature, etc.

The monthly array yield, final yield and reference yield are shown in Fig. 5. As expected, the lowest and highest monthly array yield, final yield and reference yield obtained in the month of January 2018 and March 2018 respectively. The monthly performance ratio and capacity factor of the grid connected solar PV system is shown in Fig. 6.

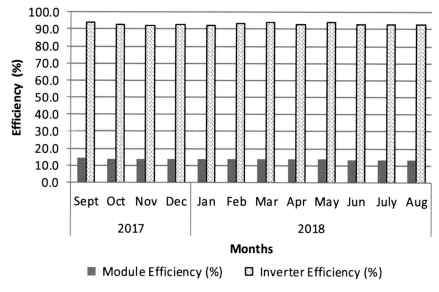

Fig. 4 Monthly solar PV module and inverter efficiencies

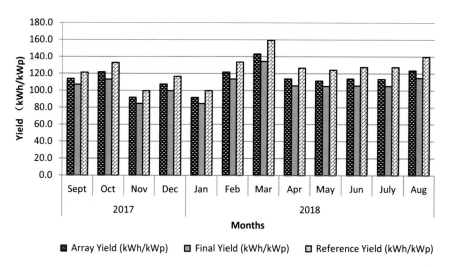

Fig. 5 Monthly array yield, final yield and reference yield

The monthly average performance ratio of this grid connected solar PV system was 84.5%. In other words, 15.5% of solar energy was not converted to AC energy due to energy loss or consumed by the equipment. Other than that, dust deposition may cause the decline of performance of a grid connected solar PV system. Performance ratio is a very useful tool to compare the theoretical and actual performance of solar

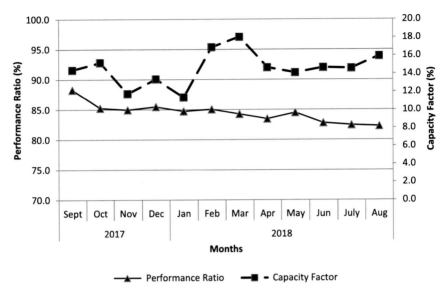

Fig. 6 Monthly performance ratio and capacity factor

PV system regardless of different location or different weather condition. All the local environmental factors at the scene is already taken into account in the calculation.

The capacity factor changes according to the AC energy productions. In March 2018, the grid connected solar PV system obtained the highest capacity factor because of highest AC energy generated, which were 18.0% and 268.6 kWh respectively. The lowest capacity factor as well as lowest AC energy generated were 11.4% and 169.3 kWh respectively in January 2018.

As discussed earlier, there is a direct relationship between AC energy output and solar irradiance. Data in February 2018 was chosen randomly to plot a graph of AC energy output versus solar irradiance as shown in Fig. 7. The intensity of solar irradiance has a significant effect on the energy production of the grid connected solar PV system. Correlation coefficient (R^2) of 0.9912 obtained from the graph indicated that there was a strong relationship between these two variables.

The equation, $P_{AC} = 1.5843H - 17.284$ demonstrated the correlation between the AC energy output (E_{AC}) and solar irradiance (H). This equation is very useful as it provides a tool to researcher to estimate the energy production of a grid connected solar PV system with only using the in-plane solar irradiance data.

Figure 8 shows the monthly average PV module temperature. The monthly average PV module temperature was between 32.0 and 37.6 °C. This small PV module temperature variation is because Malaysia has only one season and located in equatorial region with tropical weather. The average temperature of the PV module may be affected by the daily weather. When there is no rain for few days, the overall PV module temperature would rise.

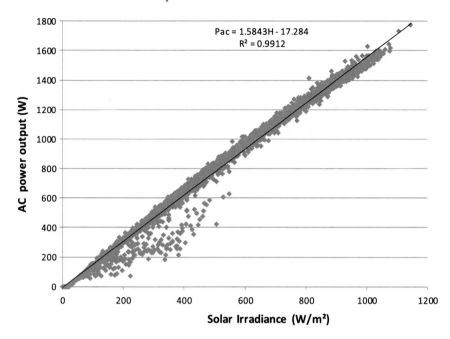

Fig. 7 AC energy output and solar irradiance in February 2018

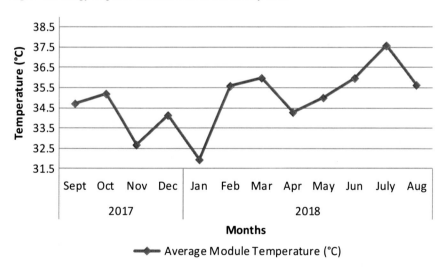

Fig. 8 Monthly average PV module temperature

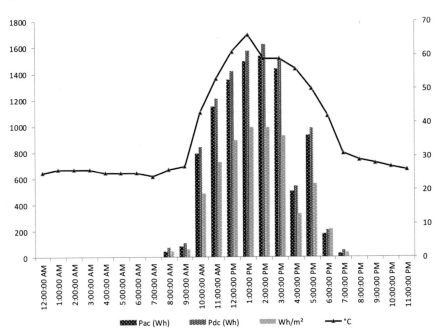

Fig. 9 DC and AC energy outputs, solar irradiation and PV module temperature on 25th of February 2018

Apart from irradiance, PV module temperature has influence on the DC and AC energy outputs where if temperature is too high the DC and AC energy outputs would reduce. Hourly DC and AC energy outputs, solar irradiation and PV module temperature on the 25th of February 2018 was chosen randomly and plotted in Fig. 9. This is a typical daily trend of the grid connected solar PV system performance. The highest solar PV module temperature was 66 °C at 1.00 pm of the day. However, the highest DC and AC energy outputs recorded that day was at 2.00 pm with PV module temperature of 59 °C that is less than the highest solar PV module temperature. Generally, the grid connected solar PV system received higher amount of solar irradiation around 10.00 am to 3.00 pm each day. Thus, it yields the highest energy output during this period. By understanding this characteristic, tilt and direction of the solar panels could be adjusted accordingly.

8 Summary

The performance of the 2-kWp grid connected solar PV system installed at Gate 3 UTP was evaluated for a period of one year, from September 2017 to August 2018. The lowest and highest recorded monthly DC and AC energy outputs were in January 2018 and March 2018 respectively. This was in agreement with the solar irradiance received by the grid connected solar PV system that was influenced by the weather. The monthly solar PV module and inverter efficiencies had deviated less than 10% from the declared efficiencies by the manufacturers. Similarly, the calculated monthly array yield, final yield and reference yield were lowest and highest in the month of January 2018 and March 2018 respectively. The monthly performance ratio and capacity factor of the grid connected solar PV system were within the accepted values for tropical climate installation. A strong direct relationship between AC energy output and solar irradiance was indicated in this study. Generally higher temperature would generate more DC and AC energy outputs. However, beyond certain temperature, DC and AC energy outputs would decline. This study concluded that the performance of the grid connected solar PV system installed at Gate 3 UTP is currently good considering this installation is only 2 years old. It is expected that the performance of solar PV system will reduce over the time as stated by manufacturer. Theoretically, the life cycle time of any solar PV system should be around 25 years.

References

1. Borhanazad H, Mekhilef S, Saidur R, Boroumandjazi G (2013) Potential application of renewable energy for rural electrification in Malaysia. Renew Energy 59:210–219
2. Hossain FM, Hasanuzzaman M, Rahim NA, Ping HW (2015) Impact of renewable energy on rural electrification in Malaysia: a review. Clean Technol Environ Policy 17(4):859–871
3. Shafie SM, Mahlia TMI, Masjuki HH, Andriyana A (2011) Current energy usage and sustainable energy in Malaysia: a review. Renew Sustain Energy Rev 15:4370–4377
4. Muhammad-Sukki F et al (2012) Solar photovoltaic in Malaysia: the way forward. Renew Sustain Energy Rev 16:5232–5244
5. de Lima LC, de Araújo Ferreira L, de Lima Morais FHB (2017) Performance analysis of a grid connected photovoltaic system in northeastern Brazil. Energy Sustain Dev 37:79–85
6. Adaramola MS, Vågnes EET (2015) Preliminary assessment of a small-scale rooftop PV-grid tied in Norwegian climatic conditions. Energy Convers Manag 90:458–465
7. Sundaram S, Babu JSC (2015) Performance evaluation and validation of 5 MWp grid connected solar photovoltaic plant in South India. Energy Convers Manag 100:429–439
8. Komoni V, Gebremedhin A, Ibrahimi N (2016) Performance evaluation of grid connected photovoltaic systems. In: Mediterranean conference on power generation, transmission, distribution and energy conversion (MedPower 2016), p 61(7)–61(7)
9. Wittkopf S, Valliappan S, Liu L, Ang KS, Cheng SCJ (2012) Analytical performance monitoring of a 142.5 kWp grid-connected rooftop BIPV system in Singapore. Renew Energy 47:9–20
10. Sarraf AK, Agarwal S, Sharma DK (2016) Performance of 1 MW photovoltaic system in Rajasthan: a case study, pp 4–8
11. Roberts JJ, Cassula AM, Celso J, Junior F (2015) Simulation and validation of photovoltaic system performance models
12. Elhadj Sidi CEB, Ndiaye ML, El Bah M, Mbodji A, Ndiaye A, Ndiaye PA (2016) Performance analysis of the first large-scale (15 MWp) grid-connected photovoltaic plant in Mauritania. Energy Convers Manag 119:411–421

Power Consumption Optimization for the Industrial Load Plant Using Improved ANFIS-Based Accelerated PSO Technique

Perumal Nallagownden, Elnazeer Ali Hamid Abdalla and Nursyarizal Mohd Nor

In recent times, energy saving has become the focus and an interesting topic for engineers and researchers. About 40–44% of the total energy is used for cooling of buildings. As cooling demand increases, the electricity consumption increases proportionally. There are numerous intelligent techniques adopted to evaluate energy usage. This chapter proposes a prediction technique to evaluate energy consumption using improved adaptive neuro-fuzzy inference system (ANFIS) model. Accelerated particle swarm optimization (APSO) was used to optimize model parameters and evaluate overall power consumption. The proposed improved model has an accuracy of 93.4%. The optimization technique reduces power consumption by 8.11% with a negligible error of 0.2. The proposed technique demonstrated the superiority of this technique compared to PSO and verified with the data obtained from Latexx Manufacturing Sdn Bhd, Perak, Malaysia.

1 Introduction

Electrical power is one of the fundamental necessities of life. With the growing population, modernity and luxury in our life today, the requirements for electrical demand is dramatically increasing. About 40% of the overall energy is consumed in building sectors [1, 2]. This energy consumption may increase more than 50% before 2030 [2, 3]. In buildings, energy consumed by cooling systems accounts for 40–44%

P. Nallagownden (✉) · E. A. H. Abdalla · N. M. Nor
Electrical and Electronic Engineering Department, Universiti Teknologi PETRONAS,
Bandar Seri Iskandar, 32610 Seri Iskandar, Perak Darul Ridzuan, Malaysia
e-mail: perumal@utp.edu.my

E. A. H. Abdalla
e-mail: neese555@gmail.com

N. M. Nor
e-mail: nursyarizal_mnor@utp.edu.my

© The Author(s), under exclusive license to Springer Nature Singapore Pte Ltd. 2020   35
S. A. A. Karim et al. (eds.), *Practical Examples of Energy Optimization Models*,
SpringerBriefs in Energy, https://doi.org/10.1007/978-981-15-2199-7_3

of the total energy consumed. The cooling systems compose of chillers, pumps, and fan systems. Chillers alone consumes about 35% of total energy consumption in buildings [4].

In the tropical climatic zones in Southeast Asia such as Malaysia, the climate is hot humid throughout the year [5]. The rising of the ambient temperature (T_{AMB}) and humidity (R_H), increase cooling demand and results in higher power consumption [6, 7]. Cooling buildings consumes 10–60% of electricity as reported in 2011 [8]. The rising of T_{AMB} and R_H not only have an impact on energy consumption, but also affects the chillers performance [9–13]. Therefore, the optimization techniques have the potential to save energy while maintaining cooling comfort in buildings.

2   Literature Review

Several optimization techniques have been used to adjust the parameters of chillers plant to reduce energy consumption. As a result, many approaches have been proposed, such as PSO, to optimize the parameters of chiller plants to reduce electricity consumption [14–23]. A clustering method by Lam et al. [24] has been used to study the effect of weather on chiller plants power consumption. It reduced electricity usage, but the models developed was based on only weather data, not including plant variables. In another study by Deng et al. [25], a heuristic algorithm based on model predictive has been used to select number of chillers by atmosphere changes. However, the atmosphere data (ambient temperature and ambient humidity) were not included in the proposed model [26]. The outdoor weather conditions have a significant effect of the cooling load if the seasons are different [27].

Artificial neural networks (ANN) and genetic algorithm (GA) have been used to evaluate chillers power consumption [28]. ANFIS has predicted a model for cooling tower, it simulated and controlled fan speed to save energy [29, 30]. Another study by ANN and random forest (RF) was applied for a data-driven to derive a model. The models predicted hourly energy consumption. The study showed that RF model did not perform efficiently due to limited parameters that have been used [31]. Also, the ANFIS model was evaluated for power consumption [32]. Xu et al., proposed PSO to optimize optimal energy consumption strategy by shifting peak load of the building [18]. Chen et al. [19] has developed a hybrid ANN and PSO algorithm for optimal energy consumption. ANN was used to train chiller load, while PSO was used to optimize chiller power consumption. In [20, 33], PSO algorithm was used to solve the problem of optimal chiller loading and save energy. Karami et al. [23] has adopted PSO on a dynamic model to find the optimum value of control parameters for the chiller plant based on chilled water and condenser water temperature set points. In this, PSO was used to control the flow rate of chilled water and cooling water. The control in flow rate saved energy by about 10.5 and 13.6% of total chiller plant energy consumption during a hot-day and moderate-day, respectively. The model fitting was 0.93 which is poor. However, PSO algorithm has a tendency to converge towards local optima or even arbitrary points rather than global optimum.

In the literature review, there is no computation model formulated that integrates with weather condition factors based on ANFIS-APSO approach. The main target of using APSO is to improve the robustness of ANFIS model. Therefore, there is a need for the proposed approach to assess the overall consumption of a chiller system with and without the impact of weather. Two model approaches were developed and optimized based on adjusted chiller temperature to assess power consumption. When outdoor is hot, electricity consumption increased to offer cooling services [6]. This chapter is organized as follows; Sect. 2 gives related literature review. Section 3 presents the measured data and constructed the proposed approach by incorporating weather factors such as ambient temperature and ambient humidity. Section 4 is dedicated for Results and Discussion including the comparison with established methods. Finally, a chapter summary will be concluded in Sect. 5.

3 The Proposed Methodology

The implementation of the proposed approach is with the simulation work and measured data. In this chapter, two objectives have been achieved and are implemented as shown in Fig. 1.

Phase (A): Computation Model:

In the literature review there is no study that implements ANFIS optimized with APSO to predict chillers temperature. The temperature prediction will be used to formulate a computation model to evaluate chillers performance. The model uses the measured data and analyses fit statistically using regression (R) and root mean square error (RMSE).

Phase (B): Optimization Approach:

After computation model has been developed in phase (A), the objective function was formulated in Phase (B). The objective function aims to reduce power consumption using Accelerated PSO. The optimization performance was evaluated by RMSE and algorithm accuracy.

3.1 Measured Data Identification

The measured (collected) data was obtained from a chiller plant at Latexx Manufacturing Bhd Sdn, Perak, Malaysia. Table 1 gives the setting and actual data. The identifying inputs data of a chiller plant which are given in Table 1. The chiller produces the chilled water with a flow rate of 29.4 kg/s at a supply temperature of 6.5 °C and the chilled water returns at 12.5 °C. To determine the cooling load

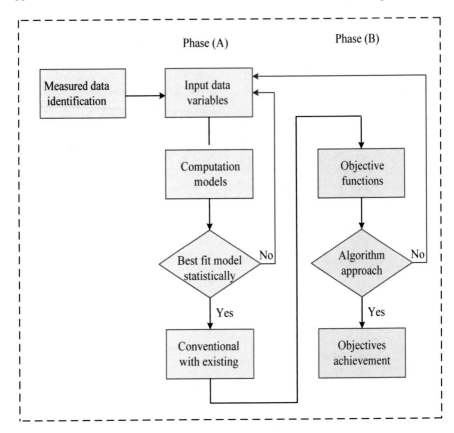

Fig. 1 Flow chart of research methodology

Table 1 The recommended and actual reading of chiller systems

No.	Variable name	Symbol	Setting	Actual
1	Supply temperature of chilled-water (°C)	T_{CHWS}	6.5	6.5–6.9
2	Return temperature of chilled-water (°C)	T_{CHWR}	12.5	8.8–16.3
3	Flow rate of chilled-water (kg/s)	M_{CHW}	29.4	11–29
4	Power consumption (kW)	P_{CHI}	140	85–165
5	Load current (A)	I_L	230	100–250
6	Voltage Supply (V)	V_S	415	406–408
7	Power factor (cos Ø)	P_{CTF}	0.85	0.82–0.84
8	Partial load ratio	L_R	1	0.3–1
9	Ambient temperature (°C)	T_{AMB}	–	24–35 [34]
10	Relative humidity (%)	R_H	–	46–100 [34]

consumption, ambient temperature and ambient humidity are taken into consideration. The range ambient temperature and humidity have been taken from Taiping, Perak, Malaysia weather history https://www.worldweatheronline.com/taiping-weather-history/perak/my.aspx. The reason of ambient temperature and humidity are that they have an impact on the cooling load performance which lead to increase/decrease power consumption.

The importance of the measured data is to build a computation model to evaluate power consumption of the chiller. The data modeling includes the variables of a chiller plant and weather factors. The atmosphere factors have an impact on the performance of the chiller, which result to increase/decrease power consumption [2].

3.2 Architecture of ANFIS-APSO Model

The ANFIS model was implemented with 4 inputs to develop the model of temperatures. Figure 2 shows the structural diagram of ANFIS used to predict return temperature of chilled water (T_{CHWR}). There are 4 inputs to predict T_{CHWR}. The inputs are (1) temperature difference of chilled water ($\Delta_{TCHW} = T_{CHWR} - T_{CHWS}$), (2) flow rate of chilled water (M_{CHW}), (3) ambient temperature (T_{AMB}), and (4) ambient humidity (R_H). All inputs have a direct effect on the return temperature.

ANFIS implemented a model with 4 layers. Layer 1 is used to enter the crisp values and transmitted to the next layer. Layer 2 utilized fuzzy memberships input. Layer 3 adopted the incoming signals of the previous layer to develop the rules after multiplying memberships. Layer 4 was used to calculate the overall outputs from incoming signals to develop the final models of T_{CHWR}.

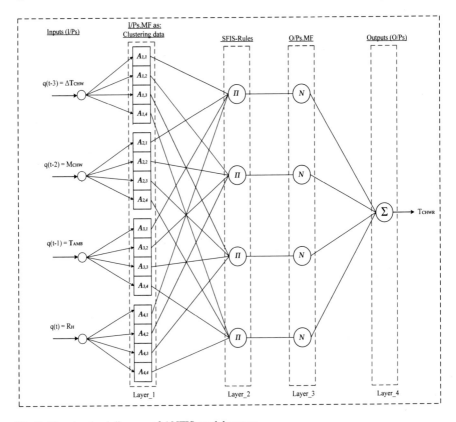

Fig. 2 The structural diagram of ANFIS model system

ANFIS was used to carry out the simulation procedure. Thus, the fuzzy inference system based on Sugeno was used to build fuzzy rules (R) [34]. It was implemented with time series 't' up to time '$t + 1$' where $t = 1, 2, ..., T$. The fuzzy rules can be formulated as;

$$
\begin{aligned}
& if\ q_j(t-3)\ is\ A_{j1}^R\ and\ q_j(t-2)\ is\ A_{j2}^R\ and \\
& q_j(t-1)\ is\ A_{j3}^R\ and\ q_j(t)\ is\ A_{j4}^R, then\ f_j(t+1) \\
& = q_j(t-3)M_{j1}^R + q_j(t-2)M_{j2}^R \\
& + q_j(t-1)M_{j3}^R + q_j(t)M_{j4}^R + M_{j5}^R
\end{aligned} \tag{1}
$$

where q_j are inputs, A_{ij}^R are antecedent's fuzzy sets, $T_{CHWRj}(t + 1)$ is the system outputs, and M_{ij}^R are premise parameters of each fuzzy rule. The premise parameters adjusted automatically with least squares algorithm. Therefore, APSO was used to improve ANFIS gradient descent in local minima [34]. The 4-layers ANFIS are defined in detail as follows:

Layer 1, each i generates membership grades of inputs. The membership degree (μ) of group (A) can be formulated based on Gaussian function as

$$\mu_{ij} A_{ij}^R (q_j(t-3)) = \exp\left\{ \frac{-(q_j(t-3) - c_i^R)}{2\sigma_i^{2R}} \right\}^2 \tag{2}$$

where μA^R is the membership functions (MF) of each rule, $q_j(t-3)$ are the inputs to the node at layer 1, c_i^R and σ_i^R are premise parameters of mean and standard deviation of MF, respectively.

Layer 2, in this, each output node determines as MF product, it computed in Eq. (3) as,

$$\mu_R = \prod_{i=1}^{i=4} \mu_{ij} A_{ij}^R (q_j(t-4-i)), \quad when \ i = 1, 2, 3, 4 \tag{3}$$

Layer 3, the results are obtained from layer 2 based on Eq. (3) represent the implication degree value as,

$$\mu_R^j = \mu_R / \sum_{j=1}^{j=4} \mu_R \tag{4}$$

Layer 4, the output of layer 3 is multiplied by a Sugeno FIS first order linear model. Thus, for each j cluster the overall output for predicting future at $(t+1)$ predicting can express as,

$$T_{CHWR}(t+1) = \sum_{R=1}^{4} \mu_R \left(q_j(t-3)M_1^R + q_j(t-2)M_2^R \right.$$
$$\left. + q_j(t-1)M_3^R + q_j(t)M_4^R \right) / \sum_{R=1}^{4} \mu_R \tag{5}$$

Equation (5) for the future prediction T_{CHWR} at $(t+1)$ can be mutated as,

$$T_{CHWR}(t+1) = \Delta T_{CHWj}(t-3)M_{ij}^1 + M_{CHWj}(t-2)N_{ij}^1$$
$$+ T_{AMBj}(t-1)Y_{ij}^1 + R_{Hj}(t)Z_{ij}^1 + O_{ij}^1 \tag{6}$$

3.3 The Performance of Computation Model

The data measured of ΔT_{CHW}, M_{CHW}, T_{AMB}, and R_H were implemented based on time-series for T_{CHWR} prediction. The simulation was carried out using ANN, ANFIS, ANFIS-APSO1, and ANFIS-APSO2. ANFIS-APSO1 has 4 inputs, whereas ANFIS-APSO has 2 inputs data. The techniques were implemented with measured data, where Table 2 gives the detailed parameters.

The fit model statistics shown is based on regression (R^2) as in Fig. 3a–d. The prediction model performance was investigated with the testing data, training data, and all data. Figure 3a shows R^2 of T_{CHWR} using ANN of 4 inputs-data. Figure 3b shows R^2 of T_{CHWR} by ANFIS of 4 inputs-data. Figure 3c shows R^2 of T_{CHWR} with ANFIS-APSO1 of 4 inputs-data. Figure 3d shows R^2 of T_{CHWR} based on ANFIS-APSO2 of 4 inputs-data.

It was observed that T_{CHWR} showed a least R^2 when it was implemented with ANN and ANFIS approach. The ANFIS-APSO1 approach with 4 inputs-data demonstrated a superiority with a good model accuracy with a regression above 98%. Whereas, ANFIS-APSO2 was implemented with only two inputs-data, showed a good regression, however testing data has a worst regression R^2.

Table 2 The simulation parameters using different techniques

Parameters	ANN	ANFIS	ANFIS-APSO1	ANFIS-APSO2
No. of inputs data	4	4	4	2
Rules number	–	16	4	4
No. of clusters	4	16	4	4
No. of nodes/neurons	20	101	47	47
No. of linear parameters	60	48	20	20
No. of nonlinear parameters	100	64	32	32
Total parameters	160	165	52	52
ra	–	–	0.52	0.52
Goal error	0.0001	0.0001	0.0001	0.0001
Decrease rate	–	–	0.85	0.85
Increase rate	–	–	1.15	1.15
α, β [34]	–	–	0.47, 0.63	0.47, 0.63
(r, r2), (rij)	–	–	0, 1	0, 1
Particles dimension	–	–	20	20
Population	–	–	50	50
Iter./epoch	250	250	250	250

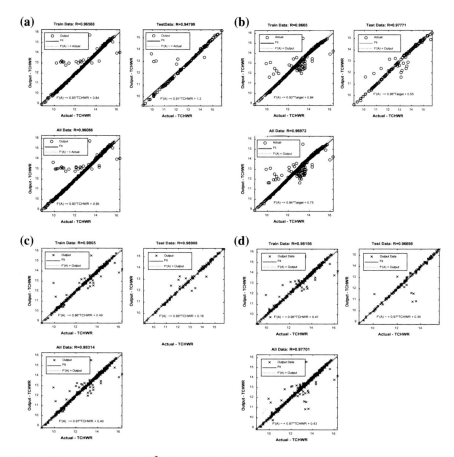

Fig. 3 The model regression R^2 with optimization techniques

3.4 The Impact of Weather Data

The prediction model was the integration of 2 weather factors (T_{AMB} and R_H). The impact of weather factors affect on the nominal values of chiller plant temperatures. The nominal values can be increased or decreased based on weather conditions. Based on that, we can operate chillers (either fully or partially). The weather conditions should be known, so that we consume power based on the need of cooling demand. The temperature T_{CHWR} model has a positive impact to assess electricity consumption of chiller plants. Also, the power consumption model can be evaluated even without considering the weather condition changes. But if weather (rainy or cloudy), and chillers are operated at full load, over-cooling can occur (not needed), and power consumption will be increased.

3.5 Optimization Model by APSO Technique

The power consumption of chillers can be expressed as,

$$P_C = \sqrt{3} * V_S I_L \cos \emptyset \tag{7}$$

Also, electricity consumption can be expressed based on T_{CHWR} and M_{CHW} as,

$$P_{CHI} = 0.76\,m \sum_{m=1}^{M} \sum_{t=1}^{T} y_I\, y_{II} + 7.7 \tag{8}$$

where m is the number of chillers, y_I and y_{II} can be expressed [34],

$$y_I = T_{CHWR}(t+1) - T_{CHWS} \tag{9}$$

$$y_{II} = M_{CHW} * L_R \tag{10}$$

The objective functions that were used to evaluate the overall performance of a chiller electricity consumption in Eq. (8) can be expressed as,

$$minimize,\ f_i = m \sum_{t=1}^{T} P_{CHI} \tag{11}$$

The objective function can be formulated with a chiller and weather variables with a computational model as expressed in,

$$T_{CHWR}(t+1) = 0.9689 * \Delta T_{CHWi}(t-3) - 0.0175$$
$$* M_{CHWi}(t-2) - 0.0087 * T_{AMBi}(t-1)$$
$$- 0.0517 * R_{Hi}(t) + 9.2318 \tag{12}$$

Equation (12) substitutes in the objective function of Eq. (11) and was implemented with ANFIS-APSO1. The objective function in Eq. (11) can be formulated with the variables of the chiller only, with a computational model as expressed in,

$$T_{CHWR}(t+1) = 0.8314 * \Delta T_{CHWi}(t-1) - 0.018 * M_{CHWi}(t) + 8.2205 \tag{13}$$

Equation (13) substituted in the objective function of Eq. (11) and was implemented with ANFIS-APSO2. Then, the output of the temperature prediction in Eqs. (12) and (13) based on ANFIS-APSO1 and ANFIS-APSO2 were used to evaluate power consumption in Eq. (11). The optimization with APSO for power consumption has been compared with standard PSO in the results section. The simulation was carried out over 24 h to minimize the chiller electricity consumption.

4 Results and Discussion

First, initialize the random swarm (population), and find the fitness of each particle, then update the velocity and position of particles at each iteration. The APSO
implementation is presented in the flow chart shown in Fig. 4.

The main parameters of APSO need to be pre-defined as listed in Table 3.

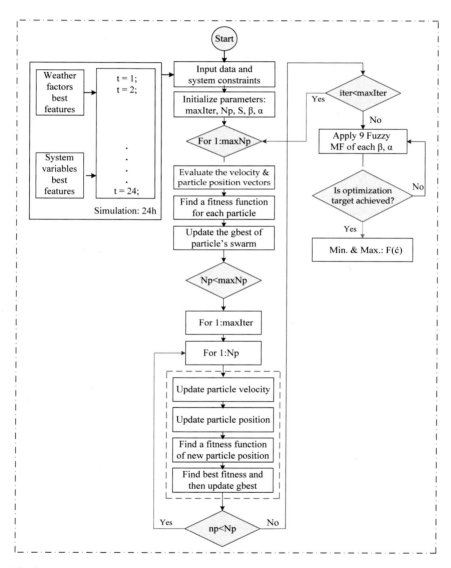

Fig. 4 The flow chart of optimization model by APSO algorithm

Table 3 The setting parameters of APSO algorithm

Parameter	Value	Parameter	Value
Swarm	50	Beta (β)	0.63
Iterations	250	No. of dimensions	2
Alpha (α)	0.47	r_{ij}	0, 1

4.1 Results

Figure 5 shows the power consumed of a 1 chiller system implemented with 3 optimization techniques over 24 h.

Table 4 gives a comparison result of the optimization techniques for two objective functions.

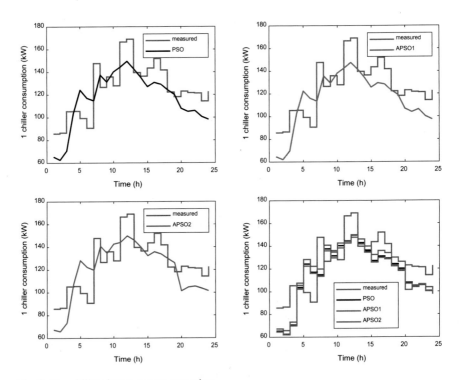

Fig. 5 One chiller plant power consumption

Table 4 The comparison results of the simulated power consumption

Power (kW)	Measured	PSO	APSO1	APSO2
Consumed	3008	2789	2764	2860
Saving %	–	7.3	**8.2**	4.9

4.2 Discussion

It was observed that in Fig. 5, the implementation of optimization by APSO1 achieved higher saving compared to the standard PSO. Even though, PSO and APSO1 were both implemented with the same data and the same variables including weather factors. Therefore, the result inferred that computation model when implemented using more variables (weather) will achieve good fit model. Then, the power consumed can be adjusted based on the need for cooling when the weather conditions are known. The objective function when implemented with optimization-based APSO2, consumed more power. It was noticed that APSO1 for 1 chiller optimization has a considerable efficient performance, while it reduced power consumption by 8.2% compared to the APSO2 and standard PSO algorithm as given in Table 4. To evaluate the performance of the proposal, a statistical analysis expression based on root mean square error (RMSE) as,

$$RMSE = \sqrt{\frac{(y - \gamma)^2}{n}} \tag{14}$$

where y and γ are actual and prediction where error ($e = y - \gamma$), and n is the number of data points. Figure 6 shows the fitness of the performance of three optimization techniques.

Here, PSO and APSO1 were carried out with their own parameters and applied with the same objective function variables. Whereas, APSO2 was implemented with the same algorithm parameters with different system variables. APSO1 includes two weather factors which are T_{AMB} and R_H, while APSO2 was not. This showed the impact of more system variables to fit the system model.

Table 5 gives the worst and best fitness based on RMSE error.

The algorithm accuracy can be evaluated based on the mean of MSE as given by RMSE in Table 6.

The optimization was implemented with a Toshiba, P6100@2.00GHz2.00 GHz (2RAM). Table 7 gives the details of the computation time for each algorithm in seconds.

The optimization technique-based APSO1 accelerated the computation time compared to the standard PSO. While APSO2 with 250 iterations reduced the computation time. This is due to a few additional variables associated and involved with the objective functions when PSO and APSO1 were used.

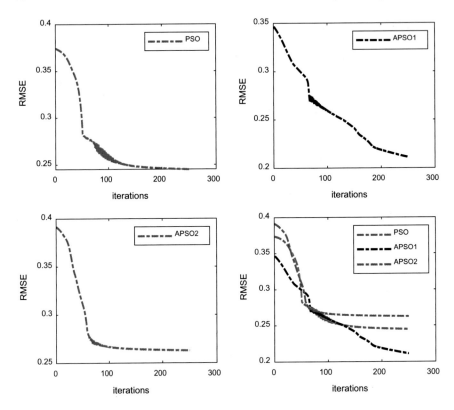

Fig. 6 The optimization performance with RMSE

Table 5 The performance of algorithms with RMSE

RMSE	Worst	Mean	Best
PSO	0.3742	0.2739	0.2452
APSO1	0.3465	0.2577	0.2103
APSO2	0.3917	0.2858	0.2632

Table 6 The accuracy of optimization performance

Model	PSO	APSO1	APSO2
Accuracy (%)	92.5	**93.4**	91.8

Table 7 The comparison based on computation time

Time	PSO	APSO1	APSO2
(s)	106.7	105.8	104.1

5 Conclusion

A computation model was developed to evaluate electricity consumption of a chiller using ANFIS-APSO technique. ANFIS was used to formulate the prediction of the temperature model, while APSO improved ANFIS in terms of model regression. Then, ANFIS-APSO was developed the prediction model of temperature with and without weather data, and this temperature was utilized to evaluate power consumption of a chiller plant. After developing temperature and power model, an objective function has been formulated. The optimization APSO was carried out and implemented on two objective functions. The outcome of the optimization model that it was implemented with APSO has a good accuracy compared to the standard PSO. For 24 h, the optimization by APSO achieved power saving of 8.11% and compared to the standard PSO of 7.28% and APSO of 4.92%, when it was implemented with plant parameters only. The optimization APSO has a good performance compared to the standard PSO in terms of the fitness by RMSE.

References

1. Shaikh PH, Nor NBM, Nallagownden P, Elamvazuthi I, Ibrahim TJR, Reviews SE (2014) A review on optimized control systems for building energy and comfort management of smart sustainable buildings. Renew Sustain Energy Rev 34:409–429
2. Khosravani HR, Castilla MDM, Berenguel M, Ruano AE, Ferreira PMJE (2016) A comparison of energy consumption prediction models based on neural networks of a bioclimatic building. Energies 9(1):57
3. Suganthi L, Samuel AAJR (2012) Energy models for demand forecasting—a review. Renew Sustain Energy Rev 16(2):1223–1240
4. Huang S, Zuo W, Sohn MDJAE (2016) Amelioration of the cooling load based chiller sequencing control. Appl Energy 168:204–215
5. Shaikh PH, Nor NBM, Sahito AA, Nallagownden P, Elamvazuthi I, Shaikh M (2017) Building energy for sustainable development in Malaysia: a review. Renew Sustain Energy Rev 75:1392–1403
6. Patterson MK (2008) The effect of data center temperature on energy efficiency. In: Proceeding of 2008 IEEE 11th intersociety conference in thermal and thermomechanical phenomena in electronic systems (ITHERM), 28–31 May 2008, 2111 NE 25th Avenue Hillsboro, Oregon, pp 1167–1174
7. Yi-Ling H, Hai-Zhen M, Guang-Tao D, Jun S (2014) Influences of urban temperature on the electricity consumption of Shanghai. Adv Clim Res 5(2):74–80
8. Chong C, Ni W, Ma L, Liu P, Li Z (2015) The use of energy in Malaysia: tracing energy flows from primary source to end use. Energies 8(4):2828–2866
9. Wang SK (2001) Handbook of air conditioning and refrigeration. ASHRAE Handbook HVAC Applications
10. Avery G (2001) Improving the efficiency of chilled water plants. ASHRAE J 43(5):14
11. Lu L, Cai W, Soh YC, Xie L, Li S (2004) HVAC system optimization—condenser water loop. Energy Convers Manag 45(4):613–630

12. Browne M, Bansal P (1998) Steady-state model of centrifugal liquid chillers: Modèle pour des refroidisseurs de liquide centrifuges en régime permanent. Int J Refrig 21(5):343–358
13. Lu L, Cai W (2001) Application of genetic algorithms for optimization of condenser water loop in HVAC systems. World-wide-web, Nanyang Technological University Nayang Press Avenue
14. Beghi A, Cecchinato L, Cosi G, Rampazzo M (2010) Two-layer control of multi-chiller systems. In: Proceeding of 2010 IEEE international conference on control applications (CCA), 8–10 Sept 2010, Yokohama, Japan, pp 1892–1897
15. Beghi A, Cecchinato L, Cosi G, Rampazzo M (2012) A PSO-based algorithm for optimal multiple chiller systems operation. Appl Therm Eng 32:31–40
16. Wei X, Xu G, Kusiak A (2014) Modeling and optimization of a chiller plant. Energy 73:898–907
17. Xu Y, Ji K, Lu Y, Yu Y, Liu W (2013) Optimal building energy management using intelligent optimization. In: Proceeding of IEEE international conference on automation science and engineering (CASE), 17–20 Aug 2013, Madison, WI, USA, pp 95–99
18. Lee K-P, Cheng T-A (2012) A simulation–optimization approach for energy efficiency of chilled water system. Energy Build 54:290–296
19. Chen C-L, Chang Y-C, Chan T-S (2014) Applying smart models for energy saving in optimal chiller loading. Energy Build 68:364–371
20. Ardakani AJ, Ardakani FF, Hosseinian SH (2008) A novel approach for optimal chiller loading using particle swarm optimization. Energy Build 40(12):2177–2187
21. Lee W-S, Lin L-C (2009) Optimal chiller loading by particle swarm algorithm for reducing energy consumption. Appl Therm Eng 29(8–9):1730–1734
22. Kusiak A, Xu G, Tang FJE (2011) Optimization of an HVAC system with a strength multi-objective particle-swarm algorithm. Energy 36(10):5935–5943
23. Karami M, Wang LJATE (2018) Particle Swarm optimization for control operation of an all-variable speed water-cooled chiller plant. Appl Therm Eng 130:962–978
24. Lam JC, Wan KK, Cheung K (2009) An analysis of climatic influences on chiller plant electricity consumption. Appl Energy 86(6):933–940
25. Deng K, Sun Y, Li S, Lu Y, Brouwer J, Mehta PG, Zhou MC, Chakraborty A (2015) Model predictive control of central chiller plant with thermal energy storage via dynamic programming and mixed-integer linear programming. IEEE Trans Autom Sci Eng 12(2):565–579
26. Alonso S, Morán A, Prada MÁ, Reguera P, Fuertes JJ, Domínguez MJE (2019) A data-driven approach for enhancing the efficiency in chiller plants: a hospital case study. Enegies 12(5):827
27. Aktacir MA, Büyükalaca O, Bulut H, Yılmaz T (2008) Influence of different outdoor design conditions on design cooling load and design capacities of air conditioning equipments. Energy Convers Manag 49(6):1766–1773
28. Chow T, Zhang G, Lin Z, Song C (2002) Global optimization of absorption chiller system by genetic algorithm and neural network. Energy Build 34(1):103–109
29. Soyguder S, Alli H (2009) Predicting of fan speed for energy saving in HVAC system based on adaptive network based fuzzy inference system. Expert Syst Appl 36(4):8631–8638
30. Hosoz M, Ertunc HM, Bulgurcu H (2011) An adaptive neuro-fuzzy inference system model for predicting the performance of a refrigeration system with a cooling tower. Expert Syst Appl 38(11):14148–14155
31. Ahmad MW, Mourshed M, Rezgui Y (2017) Trees vs Neurons: Comparison between random forest and ANN for high-resolution prediction of building energy consumption. Energy Build 147:77–89
32. Lu L, Cai W, Li S, Xie L, Soh YC (2002) Application of ANFIS in chilled water distribution process for energy savings. In: Proceeding of 2002 IEEE international conference in control and automation (ICCA). The 2002 international conference on final program and book of abstracts, 2002, pp 98–98

33. Lee W-S, Lin L-C (2009) Optimal chiller loading by particle swarm algorithm for reducing energy consumption. Appl Therm Eng 29(8):1730–1734
34. Hamid Abdalla EA, Nallagownden P, Mohd Nor NB, Romlie MF, Hassan SM (2018) An application of a novel technique for assessing the operating performance of existing cooling systems on a university campus. Energies 11(4):1–24

Cost Benefit Opportunity for End Use Segment Using Lighting Retrofit at Taylor's University

Reynato Andal Gamboa, Chockalingam Aravind Vaithilingam and Then Yih Shyong

1 Introduction

In the present days, the increasing demand for energy utilization has become a major issue due to the growth factor of greenhouse gas emissions and identified environmental problems. The increase in energy consumption is reported to be 7.5% and expected to increase in the range of 6–8% in the following years [1, 2]. Although the increase in the number of commercial buildings in Malaysia has contributed to national development, it also increases the energy demand [3]. The end users seek to improve the efficiency of energy usage as it is critical towards sustainability to alleviate the growing energy demand [4, 5]. Within the consumer sector the commercial buildings contributed 32% of energy consumption. This turns out to be the second largest user in Malaysia [6] of which air-conditioning contributed 58% of energy consumption and lightings contributed 20% of energy consumption for Malaysian buildings. Therefore, lightings (light load) and air-conditioning (power load) are the major factors of energy efficiency, which need to be focused on to optimise its usage and energy balance.

Taylor's University Lakeside Campus is considered as a commercial building with built in area of over 27 acres of space as in Fig. 1. There are five blocks A, B, C, D and E that is powered by 11 kV, 50 Hz four transformer feeding the blocks. Figure 2 shows the electrical network of the Taylor's University Energy Distribution System

R. A. Gamboa
Lyceum of the Philippines University - Batangas, Capitol Site, Batangas 4200, Philippines
e-mail: reynatoandal.gamboa@gmail.com

C. Aravind Vaithilingam (✉) · T. Y. Shyong
Faculty of Innovation and Technology, School of Engineering, Taylor's University, 47500 Subang Jaya, Malaysia
e-mail: Chockalingamaravind.vaithilingam@taylors.edu.my; aravindcv@ieee.org

T. Y. Shyong
e-mail: Yihshyong.nze@gmail.com

© The Author(s), under exclusive license to Springer Nature Singapore Pte Ltd. 2020
S. A. A. Karim et al. (eds.), *Practical Examples of Energy Optimization Models*,
SpringerBriefs in Energy, https://doi.org/10.1007/978-981-15-2199-7_4

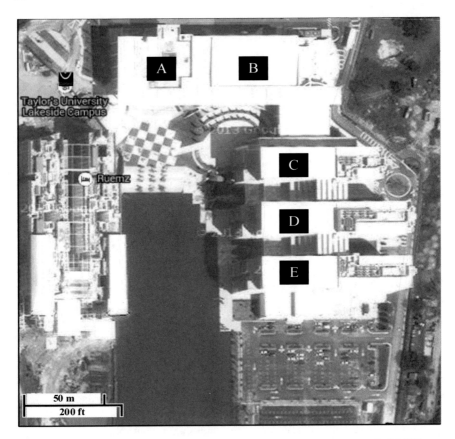

Fig. 1 Taylor's University Lakeside Campus (https://goo.gl/maps/yx5QNmwPno6gpWzTA)

(TUEDS) in place and the corresponding usage bill of Taylor's University from January 2017 to December 2017 as in Fig. 3. From the graph, the demand increases during March, July, and September. During March 2017, students are returning to university for study. During May 2017, it is seen as a rising point because June has more public holidays than other months and hence lesser energy usage. Therefore, energy usage begins to increase during July, while for September month, the energy usage is similar to March, when students are returning to study.

Air-conditioning system is considered as one of the most important loads of a commercial building, providing comfort cooling, especially in a country with a tropical climate like Malaysia. Based on MS 1525: 2014 developed by Department of Standards Malaysia, the standard temperature to be maintained in between 24 and 26 °C [7]. In this work, room temperatures are measured and compared to the standard temperature. Lighting is also considered one of the most common loads and represents 20% of energy consumption in commercial buildings [6]. It is essential as it provides a suitable, productivity, comfort and visual environment. However, it is important that the lighting systems be designed appropriately so that it provides an

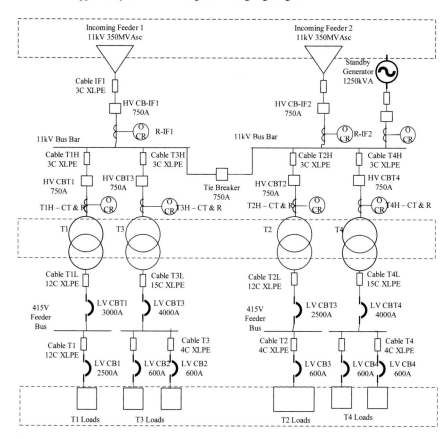

Fig. 2 Taylor's University Electrical Distribution System

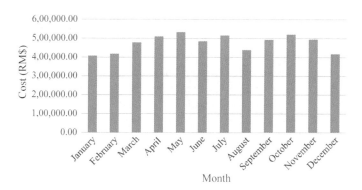

Fig. 3 Monthly bill for the year of 2017

optimal illumination level while consuming a minimum amount of energy. Therefore, lighting retrofit is used to replace inefficient lighting with efficient one [8]. This is done by minimizing input wattage of the lighting. This work focuses on minimizing the wattage by retrofitting existing lighting levels are referred to MS 1525: 2014. The corresponding recommended illumination level are 300–500, 300–400 and 100 lx [7].

2 End Use Segment Model and Approach Used

2.1 Modelling of the System

The first stage is the collection of all energy-related data required to apportion the total energy consumption into different energy end-uses. The data collected is used to build a well-grounded model of how much and where the energy is consumed. The desktop data collection is to reduce the field energy-related data when conducting data collection. This allows to understand and focus the foundation of the audited building during auditing. The data collection includes mechanical and electrical drawings, historical data of the electric bill and single-line power supply drawings. Next is the field data collection, which contributes to obtaining data which are not available during desktop data collection process with increased accuracy. Apart from that, the incoming load profile establishment is essential [9, 10]. This includes site investigation with several measuring equipment such as the lux meter and temperature measuring instrument. During the site visits, there is a method called zoning method to equally distribute the floor area in order to optimize the estimation of the load. Furthermore, visitations to Air Handling Unit (AHU) rooms is conducted to understand the number and location of AHUs in the campus. After field data collection, cross-checking of load demand data is performed. This process includes estimating building's total and end-use energy consumption. However, due to several hypotheses, the accuracy may vary depending on the load measured. During this process, data loggers are used to measure the daily load used. After acquiring the data required, the end-use load apportioning is conducted. The data collected from power logger such as energy consumption is verified with the monthly electricity bill and estimate the building monthly and end-use energy consumptions. Also, it is ensured that the energy consumption is taken into account when the total building and end-use energy consumption are calculated. Next, data analysis is performed.

2.2 Building Energy Retrofit

From the analysis, we identify energy wastage and possibility of load shedding. This process is also known as building energy retrofit. Then, cost benefit analysis is analyzed to identify the return on investment of this energy audit. From the cost benefit analysis, cost-saving opportunities are identified. Figure 4 shows the detailed flowchart of lighting retrofit and temperature measurement. This includes the latest lighting information such as wattage, location and quantity of lighting. The information acquired is the most recent to avoid incorrect justification. Also, the temperature settings are also acquired from the maintainers. Using the zoning method is used to equally distribute the floor area in order to optimize the estimation of the load. Next, the illumination level and temperature are measured using lux meter and temperature instrument. The parameters measured are checked with the data collected from the maintainers. After check and balance, the measured parameters are compared to MS 1525: 2014 [7]. This helps to identify the inadequate parameters and opportunities. Therefore, by filtering the data measured, the inadequate Lux levels are highlighted as potential lighting retrofit. After data filtering, the lighting retrofit is performed by replacing fluorescent lamps with LEDs. Before measuring the Lux

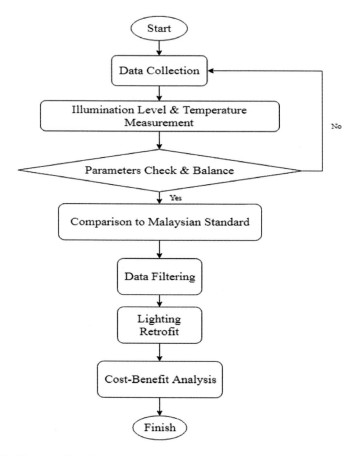

Fig. 4 Lighting retrofit and temperature measurement

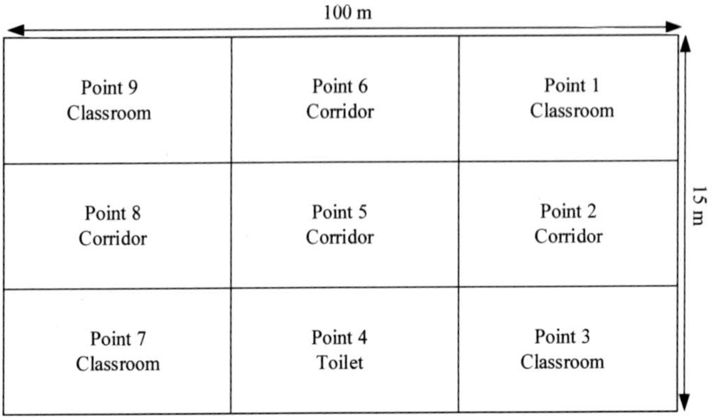

Fig. 5 Nine-point zones for each floor

levels of each floor, zoning strategy is prepared to distribute the zones evenly. This increase the accuracy and optimize the comfort of the occupant [11]. In our case, we equally distribute the area for each floor and divide them into 9 zones. Figure 5 shows the distributed points. Each of the floor has an area of estimated 100 m × 15 m. Therefore, each point has an area of 33.3 m × 5 m which equals to 166.5 m^2.

Since points 2, 5 and 8 are corridors, the recommended illumination level is 100 lx. While for points 1, 3, 4, 6, 7 and 8 are either classrooms or offices. Thus, the recommended illumination levels are 300–500 lx or 300–400 lx. Apart from measuring the area of the floor, the selection of the types of lighting is also important. Therefore, it is essential to understand the difference of lightings.

2.3 Choice of Lighting Retrofit

Table 1 shows the comparison between Light Emitting Diodes (LEDs), Compact Fluorescent Light (CFL) and incandescent light. Each type of lighting has their

Table 1 Comparison of CFL, incandescent and LED

	CFL	Incandescent	LEDs
Average life span (h)	8000	1200	50,000
Wattage (W)	13–15	60	6–8
Efficiency	Medium	Low	High
Cost per bulb	RM 7	RM 3	RM 12
Turn on instantly	No	Some	Yes
Contains mercury	Yes	No	No
RoHs compliant	No	Yes	Yes
Annual operating cost	Medium	High	Low

own drawback. Incandescent lightings have high power consumption, therefore high operating cost. CFL are not environment-friendly and takes time to switch on. LEDs have a high cost per bulb, thus initial cost of a lighting retrofit is high. However, the average life span of LEDs is 6 times of CFL and 40 times of incandescent lighting. Therefore, by using cost-benefit analysis, payback period is determined to identify if lighting retrofit help reduce power consumption and cost of electric bills. Energy Efficiency Retrofit (EER) of existing buildings is a key program for improving building energy efficiency in northern regions of China is presented in [12]. A framework to conduct a case study of a retrofit empirically examines its economic sustainability. The selection of retrofit materials also greatly influences the economic outcomes. With the excessive energy consumption worldwide, the demand for saving strategies increases [13, 14].

3   Data Capture and Cost Effective Analysis

The analysis is performed by extracting the power logging data at one-day basis so that the data is matched based on the load distribution in Table 2.

Therefore, the data extracted on a Thursday which is the common day for every power logging data as in Fig. 6. It is observed that T1 (Fig. 6a) has consumed more energy amongst the four transformers. The maximum power consumption reached to approximately 711 kW at 1:25 pm. This T1 solely supplying electrical power to the two chillers. This proves that the air-conditioning system is the top contributor amongst the other loads. It is observed that minimum power consumption is 2.04 kW. This is because the power consumption for chillers are cut off once it reaches mid-night, leaving only minor electrical equipment running such as the control system equipment. Therefore, it consumes the minimum power during dawn compared to peak period i.e. 12 p.m. noon. In addition, the chillers are switched on once it reached 6:55 a.m. For T2 (Fig. 6b), the maximum power consumed is 321 kW at 11:52 a.m. The primary loads are the lightings and mixed loads for Block A and B, primarily office space. For T3 (Fig. 6c) the maximum power consumed is 315 kW at 2:43 p.m. The primary loads are the Variable Refrigerant Flow (VRF) for Block C, D and E

Table 2 Distribution of loads in Taylor's University

Transformer	Primary load	Secondary load
T1	Chillers	
T2	Lightings and mixed loads for Block A and B	Lifts for Block A and B
T3	VRF for Block C, D and E	Lifts, fire alarm system and lightings and mixed loads for Block C, D and E
T4	Lightings and mixed loads for Block C, D and E	

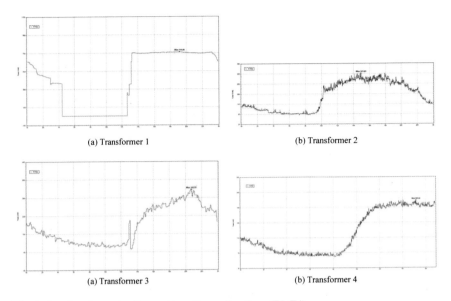

(a) Transformer 1 (b) Transformer 2

(a) Transformer 3 (b) Transformer 4

Fig. 6 Data logging at the different transformer locations (T1–T4)

(more of stochastic load points). For T4, the maximum power consumed is 208 kW at 2:38 p.m. The primary loads for transformer 4 are lightings and mixed load for Block C, D and E (also heavy load units). The power factor for the four transformers are all maintained at the level of 0.85 above.

4 Cost Effective Lighting Retrofit Opportunities

It is determined that the potential lighting retrofit are 2 zones of corridor, 2 zones of staff offices and 20 zones of classrooms. Table 3 shows the number and types of lamp existing in corridor and classroom.

Since the area of 1 point is estimated to be 166.5 m^2, the lumen output is 900 lumens, the lux required for corridor is 100 and utilization factor and maintenance factor are 1 and 0.8. The number of LEDs required to accommodate the illumination level is determined using the following formula:

Table 3 Options at each zone for corridor and classroom

Types of lamp	Location	
	Corridor	Classroom
Anabess corridor, 36 W	13	
Fluorescent tube, 18 W		3
Fluorescent tube, 36 W		16

$$No.\, of\, LEDs\, required$$
$$= \frac{(Stamdard\, Lux\, -\, Measured\, Lux)\, \times\, Area}{Lumen\, Output\, \times\, Utilization\, Factor\, \times\, Maintenace\, Factor} \quad (1)$$

Based on the scope of the proposal not all of the lightings to be replaced. Thus, the number of LEDs required is calculated by finding the difference in lux for each zone. From there, the number of LEDs is determined. Therefore, the number of LEDs required for corridor is approximately 10 LEDs. Thus, this is the total number of LEDs required for retrofitting for 2 zones of corridor area which equal to 20. The cost of each 9 W T8 LED is estimated to be RM 40 each. Therefore, the estimated total cost for retrofitting LED in corridor areas is RM 800. The load reduction per LED is also calculated. The load reduction per LED for replacing 36 W Anabess corridor (high frequency electronic ballast lighting) is 27 W. Since, the load is reduced by 27 W, the estimated energy savings are calculated by considering 8 operating hours per day and 251 operating days per year considering weekends and public holidays with 114 days not in full operating load (primarily due to semester break/holidays). The total load reduction is calculated by multiplying load reduction and total number of lamps. From there, the annual energy saving is calculated by multiplying the total load reduction with daily hours of usage and annual operating days. The estimated cost saving is calculated using the commercial building tariff rate which is RM 0.435 sen/kWh and for each kW of maximum demand is charged RM 30.3/kW. Therefore, it is determined that the total energy saving per year for retrofitting corridor areas is RM 668 and the payback period is 1.19 years. The cost-benefit analysis of potential lighting retrofits for corridors, office and classroom are shown in Table 4. From the cost-benefit analysis, replacing fluorescent tube 36 W to T8 LED 9 W could have an annual cost saving of RM 16,669. In addition, the payback period for the retrofitting is only 1.1 years. On the other hand, retrofitting fluorescent tube 18 W and PLC pin type 18 W only have annual cost saving of RM 2672 and RM 400 respectively. And the drawback of retrofitting these two types is that the payback period is too long which 3.59 years. Besides that, retrofitting could also be considered as their payback period of 1.19 years and save RM 668 and RM 1469 annual for corridor and office areas respectively.

5 Summary

The lighting retrofit is also calculated based on the lux level measured compared to the standard lux level. Each type of lamp was recorded along with the quantity and venue. The number of LEDs required is calculated and the lux required is the difference between the measured and standard lux level. This is done because to maintain the lux level all of the lightings need not be retrofitted considering optimal setting. Therefore, cost-benefit analysis is calculated based on each zone type because each zone consists of different number and types of lighting. Thus, it is found that

Table 4 Types and quantity of lamp at each zone for corridor and classroom

No.	Corridor		Classroom (18 W)		Classroom (36 W)		Office (36 W)		Office (18 W)	
	LEDs required	Anabess corridor	LEDs required	Fluorescent tube	LEDs required	Fluorescent tube	LEDs required	Anabess corridor	LEDs required	PLC pin type
Corridor for 2 zones	20	26	240	60	460	320	44	24	36	10
Load reduction (W)	27		9		27		27		9	
Lamps to be retrofitted	20		240		460		44		36	
Total load reduction (kW)	0.54		2.16		12.42		1.188		0.324	
Annual energy savings (kWh)	1084.32		4337.28		24,939.36		2385.5		650.59	
Cost per T8 LED	RM 40		RM 40		RM 40		RM 40		RM 40	
Number of T8 LED	20		240		460		44		36	
Lighting retrofit cost	RM 800		RM 9600		RM 18,400		RM 1760		RM 1440	
Cost saving per year	RM 668		RM 2672.1		RM 16,669.5		RM 1469.65		RM 400.8	

(continued)

Table 4 (continued)

No.	Corridor		Classroom (18 W)		Classroom (36 W)		Office (36 W)		Office (18 W)	
	LEDs required	Anabess corridor	LEDs required	Fluorescent tube	LEDs required	Fluorescent tube	LEDs required	Anabess corridor	LEDs required	PLC pin type
Total savings	RM 21,888.05									
Payback period (years)	1.19		3.59		1.1		1.19		3.59	

lighting retrofit saves up to RM 21,888.05 annually. The total payback period is approximately 10 years. Apart from the lightings, the room temperature is measured in each room due to the retrofitting considering the ambience. The study results are presented and subsequently research to integrate to the existing system is in the proposal stage by the facilities management side.

Acknowledgements The work is done for the grant number TRGS/MFS/1/2017SOE/007 funded by the Taylor's University.

References

1. Qureshi A, Weber R, Balakrishnan H, Guttag J, Maggs B (2009) Cutting the electric bill for internet-scale systems. ACM SIGCOMM Comput Commun Rev Newsl 39(4):123–134
2. Shaikh PH, Nor NBM, Sahito AA, Nallagownden P, Elamvazuthi I, Shaikh MS (2017) Building energy for sustainable development in Malaysia: a review. Renew Sustain Energy Rev 75:1392–1403
3. Hassan JS, Zin RM, Abd Majid MZ, Balubaid S, Hainin MR (2014) Building energy consumption in Malaysia: an overview. J Teknologi 70(7):33–38
4. Albadi M, El-Saadany E (2008) A summary of demand response in electricity market. Electr Power Syst Res 78:1989–1996. https://doi.org/10.1016/j.epsr.2008.04.002
5. Escrivá-Escrivá G, Santamaria-Orts O, Mugarra-Llopis F (2012) Continuous assessment of energy efficiency in commercial buildings using energy rating factors. Energy Build 49:78–84
6. Sadrzadehrafiei S, Mat KSS, Lim C (2011) Energy consumption and energy saving in Malaysian office buildings. Model Methods Appl Sci 75:1392–1403
7. Department of Standard Malaysi. MS 1525: 2014 Code of Practice on Energy Efficiency and Use of Renewable Energy for Non-Residential Buildings, 2014. Available: https://www.scribd.com/doc/297929846/MS-1525-2014. Accessed 04 June 2018
8. Mahlia TMI, Said MFM, Masjuki HH, Tamjis MR (2005) Cost-benefit analysis and emission reduction of lighting retrofits in residential sector. Energy Build 37(6):573–578
9. Aravind CV, Kannan R, Daniel I, Suresh A, Krishnan S (2018) Load profiling and optimizing energy management systems towards green building index (Chap. 3). In: Sustainable electrical power resources through energy optimization and future engineering. Springer, pp 25–35
10. Silsbee CH, Kostopoulos S (1999) Load profiling: a california application. J Regul Econ 15:199. https://doi.org/10.1023/A:1008085912236
11. Soori PK, Vishwas M (2013) Lighting control strategy for energy efficient office lighting system design. Energy Build 66:329–337
12. Yuming L, Tingting L, Sudong Y, Liu Y (2018) Cost-benefit analysis for energy efficiency retrofit of existing buildings: a case study in China. J Clean Prod 177:493–506
13. El-Darwish I, Gomaa M (2017) Retrofitting strategy for building envelopes to achieve energy efficiency. Alex Eng J 56(4):579–589
14. Ma Z, Cooper P, Daly D, Ledo L (2012) Existing building retrofits: methodology and state-of-the-art. Energy Build 55:889–902

A Study of Electrical Field Stress Issues in Commercial Power MOSFET for Harsh Environment Applications

Erman Azwan Yahya and Ramani Kannan

The inherent characteristics of power Metal Oxide Semiconductor Field Effect Transistor (MOSFET) for high switching speed operation at low power application makes the device very important especially in harsh environment space application. However, in the space application the device needs to withstand with radiation ambiance. This radiation source divided into two parts: particle radiation and photon radiation which are generally differentiated based on their mass and energy level. Radiation gives a significant impact to the performance of power MOSFET because of the passes of radiation ions through the device. Then, high electrical field stress arising at Junction Field Effect Transistor (JFET) region in the device's structure. Previous study shows that the power MOSFET suffers from the radiation pollution especially Single Event Effect (SEE) phenomena. SEE is an instantaneous phenomenon that happens for a few microseconds. In this chapter, a study on the radiation effect toward commercial power MOSFET by using Sentaurus Synopsys software is presented. The simulation results reveal that the electrical field stress phenomena significantly affect the performance of the commercial power MOSFET.

1 Introduction

In the 1970s, power MOSFET invented for increasing the capability of electronic device to perform high speed switching operation and handle power spikes in the inductive switching circuit. The power MOSFET initiated with the vertical structure which allows the current flow out of the drain-substrate region upward through the drain epitaxial region, then laterally across the channel to the source. This structure

E. A. Yahya · R. Kannan (✉)
Electrical and Electronic Engineering Department, Universiti Teknologi PETRONAS,
Bandar Seri Iskandar, 32610 Seri Iskandar, Perak Darul Ridzuan, Malaysia
e-mail: ramani.kannan@utp.edu.my

© The Author(s), under exclusive license to Springer Nature Singapore Pte Ltd. 2020
S. A. A. Karim et al. (eds.), *Practical Examples of Energy Optimization Models*,
SpringerBriefs in Energy, https://doi.org/10.1007/978-981-15-2199-7_5

provides a vast depletion region in the epitaxial layer as to block the high drain-source voltage when biased in the off state [1–3].

Power MOSFET offered many advantages over other semiconductor devices especially for harsh environment application including space mission. In spacecraft, this device used as power electronic supply to serve as the shunt regulator to maintain steady bus voltage, and the switching device in DC converter for battery charge assemblies [4]. Unfortunately, the application of this device in a critical environment potentially creates device breakdown thus ultimately collapse the whole system.

This circumstance has encouraged the development of reliability study on power MOSFET as a mitigation strategy to encounter reliability issue. Power MOSFET is a cornerstone to space mission application, making their reliability within the harsh space radiation environment requisite to mission success. Eventually, the motivation of this present study is to investigate the effect of radiation on commercial power MOSFET by using Sentaurus Synopsys software simulation [5–7].

This chapter is organized as follows. In Sect. 1, we give some basic introduction of power MOSFET and its application in space environment. Section 2, provide the background of radiation environment and radiation phenomena in details. Section 3 is devoted to the methodology of software simulation process by using process simulation and device simulation in Sentaurus Synopsys software. Section 4 is dedicated for results and discussion including the comparison between different types of radiation source. Summary will be given in the final section.

2 The Topology of Space Radiation Environment

In space environment, all the natural element are energetic ion combining with electron, proton, photons and other heavier ion that form radiation environment. All these particles are produced from solar particle event and galactic cosmic ray during supernova explosions. Most of the particles are trapped in the planetary magnetic field, then formed a radiation belt at planetary magnetosphere [8, 9]. This trapped particle became top of the topic to prove its sustainability in all kind of ambient environment.

For the electronic device such as power MOSFET, this particle increases the degradation in device performance and potentially causes catastrophic failure. Hence, understanding the ideal withstanding level for pollution in space is crucial to validate its sustainability. This electronic system is generally affected by the type of radiation environment. There are several types of radiation which are; trapped radiation belts, deep space cosmic rays, and transient radiation from solar flares [10–12].

I. Trapped Radiation Belt

Van Allen radiation belt is a region where most of the energetic charged particle from solar wind and cosmic rays trapped in the Earth's magnetic field. The belt divided into inner and outer belt based on the region covered above the earth's surface. The inner radiation belt contains very energetic protons produced from the collision between cosmic ray ion and atoms of the atmosphere. Other particles such as ions are less and took several years to accumulate. For the outer radiation belt, they contain less energy of ions and electrons, but the number of fluctuates increase when magnetic storm injects new particles from the tail of the magnetosphere, then gradually falling off again. In the outer radiation belt, most of the ion are protons and an alpha particle [10–12].

II. Deep Space Cosmic Rays

There are three main resources of the cosmic ray's radiation phenomena which are from the sun, galactic region of our galaxy, and extragalactic region consists of atoms that rain down from the sun and blaze at the speed of light. This atom was the most prominent failure mechanism for all the electronic application in space application. This atom can be divided into three types which are protons consist about 87%, helium about 12 and 1% for other heavier ions and electrons. Generally, the range of energy in deep space cosmic rays are from few tens of MeV to energies as high as 1018 MeV [10–12].

III. Solar Flares

A solar flare is the phenomena where the magnetic energy that has built up in the solar atmosphere is starting to be released. When the magnetic energy release, all the ions in the solar atmosphere for example electrons, protons, and heavy nuclei are heated and accelerate. Typically, the solar flare has three stage. First, the precursor stage where the release of magnetic energy is triggered. During this stage, a soft X-ray emission detected. The second stage is the impulsive stage where the protons and electrons accelerated to energies exceeding 1 MeV. During this stage, radio waves, hard X-rays, and gamma rays emitted. The last stage is the decay stage, where X-rays gradual build-up and decay of soft X-ray within a few seconds to one hour [10–12].

IV. Single Event Effect (SEE) Phenomena

A single event effect (SEE) is the phenomena when the passage of a single energetic particle through the electronic device and make the device failure. It happens in very short duration within a few microseconds at an appropriate energy level of a single energetic particle that enough to makes electronic devices degradation and failure. In space mission, heavy ions and protons were the main factors of SEE. During radiation, the energetic particle will form a track inside the device, ionized charge along its track, and generate a new electron-hole pair inside the device by losing their energy. The accumulation of generated new electron-hole pair gives electrical stress in the device and will degrade the performance of the device. Generally, SEE is a catastrophic failure for the electronic devices that happened in harsh environment application and have two type which are Single Event Burnout (SEB) and Single Event Gate Rupture (SEGR) [13–15].

SEB is the phenomena that will fluctuate the electrical behaviour of the device. SEB happened when single heavy ion particle penetrated on the device. This heavy ion particle create a bridge (short circuit) between the drain and source terminal.

Single heavy ion generates a new-electron hole pair by depositing their high energy in the pathway. The accumulation of the new electron-hole pair in this region caused numbers of internal short circuit from the source terminal to drain terminal and make high saturation of electrical field density inside the device. Then, the value of drain to source current will drastically increase and the device will burnout. In addition, SEB gives significant effect to p-channel MOSFET because the impact of ionization rate of hole is high compare to electron due to the different of the magnitude of avalanche generated current [16–18].

Second catastrophic failure for the electronic device is SEGR. SEGR can be categorized as permanent damage to the device because it will vandalize the structure of gate oxide. Hence, the main purpose of the gate as to control flowing of current from source to drain is interrupted. Besides, for the SEGR, the angle of radiation penetration also is affected. The energetic particle that penetrated perpendicularly to the gate structure will enhance the dielectric breakdown. When a positive voltage applied to the gate, these particle will accumulate at the Si/SiO_2 interface and form a conduction bridge between gate and drain. Then, the high electrical stress form at the gate oxide and localizing gate rupture indirectly make dielectric breakdown. In term of electrical characteristics, SEGR give a significant impact to device by shifting the threshold voltage and increase drain-source voltage because the change in the drain-source resistance [19, 20].

3   Software Simulation

In Sentaurus Synopsys software, two main features are used for designing process which are process simulation and device simulation as shown in Fig. 1. Before start with the designing process, the Sentaurus workbench must be setup. The workbench creates the environment of the project and can set the numbers of simulation at one time.

For process simulation, two option provided for this process which are Sentaurus Device Editor (SDE) and Sentaurus ligament. SDE is preferable because it is easy to modify the device structure and connected with Graphical User Interface (GUI) to input process step. However, for the mesh, it is complex compare to Sentaurus ligament because the mesh need to organize by command file and GUI.

For the Sentaurus ligament, all the industrial process can be undergoing in details and it also easy for meshing. However, to make optimization for the device geometrical structure is very difficult. For this method, if any modification in doping, size of region and etc., the preliminary result for the device behaviour and characteristic between each of interface is a must and a lot of work need to be done [21, 22]. For this research work, SDE is selected because of the suitability with the research core area and focus.

After finished with process simulation, the device will proceed with the device simulation by using the Sentaurus Device (SDevice) tool. SDevice will utilized all the information from the SDE tool regarding the structure grid file and parameter

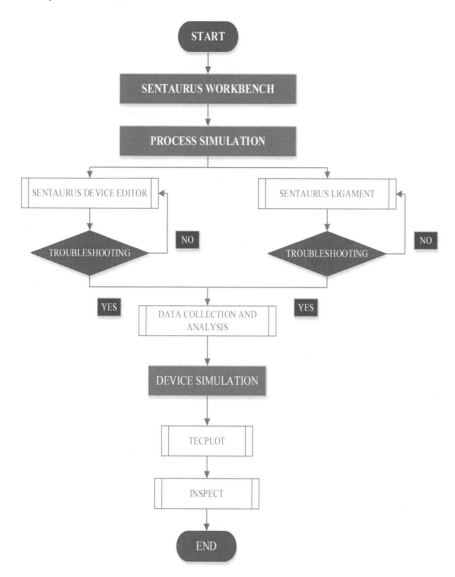

Fig. 1 Technical flowchart for the software simulation

files for the device. At this time, SDevice can start device simulation based on physic model listed in command. Under this step, two features are used to plotting the result. The result will be viewed by using Tecplot and Inspect. Tecplot tool will display the image of structure patterns and Inspect will plot the graph of electrode current and voltage. Device simulation basically to study the structure of the designed device and to analyse the I-V characteristic of Power MOSFET [20, 21].

Fig. 2 The command
structure for radiation setup
in physics section

```
Radiation {
Dose = <float> | DoseRate =
<float>
DoseTime = (<float>,<float>)
DoseTSigma = <float>

}
```

3.1 Gamma Ray Radiation

For the photons radiation simulation, gamma-ray has been selected as the radiation source in the SDevice tool. All the specification of the radiation such as dose time, energy level and dose rate were set in the physics section in the command and activated by specifying the keyword 'Radiation' as shown in Fig. 2.

From Fig. 2, the first parameter need to consider is the dose rate. The value of dose rate was calculated by using the formula in Eq. 1 below and the unit rad/s.

$$G_r = g_o D \times Y(F) \tag{1}$$

Second parameter is the dose time or the time of radiation penetration on the electronic devices. The magnitude of radiation period need to be decided in the optional Dose Time (s). Next parameter need to concern is the DoseTSigma (s). The DoseTSigma will determine the standard deviation of a Gaussian rise and fall of the radiation exposure by considering the Dose Time (s). After all, the total radiation exposure over the prescribed Dose Time Interval can be set in Dose Rate (in rad). To plot the generation rate due to gamma radiation, specify 'Radiation Generation' in the plot section [21, 22].

$$Y(F) = \left(\frac{F + E_o}{F + E_1}\right)^m \tag{2}$$

D is the dose rate, g_0 is the electron-hole pairs generation rate while E_0, E_1, and m are constants.

3.2 Particle Radiation

For the particle radiation, single heavy ion radiation model used in the software. Figure 3 shows the analogy of radiation incident in the electronic device. Based on the observation, the radiation particle create a pathway inside the device structure due to accumulation of new electron hole pairs that generated from high-energy particle by depositing their energy. This scenario will cause a high current density in the device and indirectly fluctuated the I-V characteristic of the device. There are

Fig. 3 Heavy Ion model

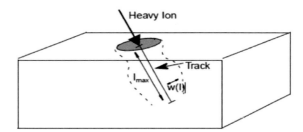

several factor that will influence the generation of the new electron hole pair such as; energy level of ion, type of ion, angle of ion's penetration and lastly the relation between the linear energy transfer (LET) and the number of pairs created.

As aforementioned, for the particle radiation, single heavy ion is used as the radiation source for the reliability test for power MOSFET. Figure 3 shows a simple model for the heavy ion impinging process. There are few parameters need to be determined for this model such as; the length of the track, l, the width, w and the temporal variations of the generation rate, $T(t)$. The generation rate caused by heavy ion is computed by Eq. 3.

$$G(l, w, t) = G_{LET}(l)R(w, l)T(t) \tag{3}$$

The linear energy transfer generation density, $G_{LET}(l)$ is the amount of energy required to transfer the ionizing particle to the material traversed per unit distance. It describes the action of radiation into matter. Besides, the Gaussian function, $T(t)$ is shown in Eq. 4 as below.

$$T(t) = \frac{2e^{\left[-\left(\frac{t-t_o}{\sqrt{2}\times S_{hi}}\right)^2\right]}}{\sqrt{2}S_{hi}\sqrt{\pi\left(1 + erf\left(\frac{t_o}{\sqrt{2}\times S_{hi}}\right)\right)}} \tag{4}$$

where t_o is the moment of the heavy ion, and S_{hi} is the characteristic value of the Gaussian. The spatial distribution $R(w, l)$ can be defined as an exponential function (default) as shown in Eq. 5 below.

$$R(w, l) = e^{\left(-\frac{w}{w_t(l)}\right)} \tag{5}$$

The radius, w defined as the perpendicular distance from the track. The characteristic distance is defined as W_{t_hi} in the Heavy Ion statement and can be a function of the length, l [21, 22]. After finish modelling, all the specification need to put in the software SDevice under physic modelling part.

4   Results and Discussion

This section discusses about the result from simulation work by using Sentaurus Synopsys software. The result presented is the explanation for gamma ray radiation and single heavy ion radiation effect on the power MOSFET by using two variables which are; Time-based study (Dose Time) and Linear Energy Transfer (LET) based study (Dose Rate).

4.1   Gamma Ray

As aforementioned before, there are two variable used for the simulation which are dose rate and the dose time. Figure 4 shows the device structure for the electrical field in the device before and after the radiation with the gamma ray radiation at a dose rate of 25 MeV when 10 V of voltage applied to the gate. As can be seen in the both figures, there are slightly different between both structures especially in the JFET region. The density of electrical field is slightly higher in the radiated device compare to the virgin device due to the accumulation of secondary electron-hole pair in the Si/SiO$_2$ interface. The radiated device may have higher electrical field if higher gate voltage applied.

In other aspect, gamma ray radiation also affected the electrical characteristic of the device. Figure 5 shows the graph of drain current versus drain voltage for the time based study during on-state region. From the result, the value of drain current is high for the radiated device compare to the virgin device when 10 V applied to the gate terminal. The radiation exposure for gamma ray is set to be only 1 ms. This duration of time is significant and enough to cause the disturbance in device performance.

For the dose rate study, the dose time value is constant for all the dose rate level which is 1 ms. Figure 6 presented the graph of drain current versus drain voltage during off state region where there is no voltage applied to the gate electrode for the dose rate of 0 MeV (Virgin), 25, 50, 75, and 100 MeV. From the obtained result, there are some fluctuation of electrical characteristic between each of the dose level.

(a)   **(b)**

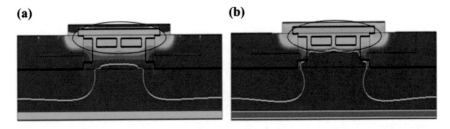

Fig. 4 Electrical field on the device for photon radiation; **a** the virgin device, **b** the device after the radiation

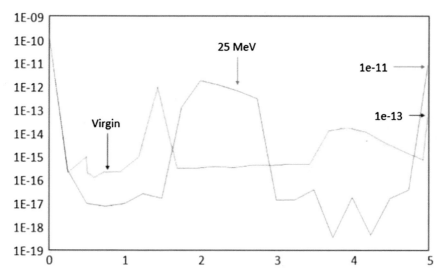

Fig. 5 Graph of I_{DS} versus V_{DS} for time-based study for photon radiation

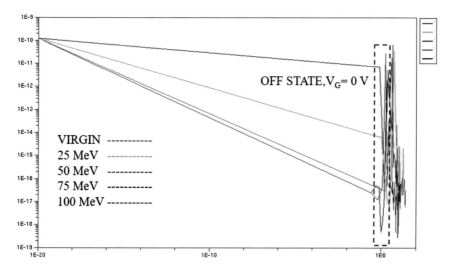

Fig. 6 Graph of I_{DS} versus V_{DS} for dose rate (LET-based study) for photon radiation

The value of drain current is varied with the dose rate level, the higher value of the dose rate will result the higher shifting in I_{DS} value during the off-state region $V_G = 0$ V.

(a)   **(b)**

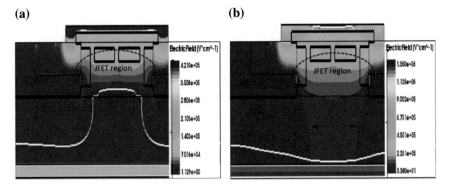

Fig. 7 Electrical field on the device for particle radiation; **a** the virgin device, **b** the device after the radiation

4.2 Particle Radiation

For the particle radiation, single heavy ion radiation particle is used in the simulation. Generally, particle radiation will cause a great impact to the device performance because they have charge and mass. Figure 7 shows the device structure before and after the radiation with the single heavy ion radiation source for the electrical field behaviour at gate voltage of 10 V. For this condition, the period of heavy ion penetration on the device used is 1 ms and 25 MeV of LET The result shows, the device that experienced with the radiation has higher electrical field in the drain-source region. This phenomena happened due to the generation of new electron-hole pair that produce by radiation particle. This new generation electron-hole pair will make the density of electron and hole increase drastically then make a field saturation in the region. In critical case, the device will breakdown or burnout due to thermal issues.

In term of the electrical performance, single heavy ion also bring a huge impact to the device. First study focus on the effect of the dose time, where the dose rate is constant at 25 MeV. Figure 8 shows the graph of drain current and drain voltage for the time based study during off state region. From the graph, the value of the drain current is changing drastically once device is exposed to the radiation either short or long radiation for dose time. This trend prove that, the SEE phenomena can occur at very short period if the dose level is significant to cause the device degradation.

Moreover, for the dose rate variable the dose time value is constant which is set to be 1 ms. As can be observed from the result in Fig. 9, the electrical performance of device is changing and varied with the dose rate amount. From the result, the higher dose rate will cause lower drain current value when the device is operating in off state region.

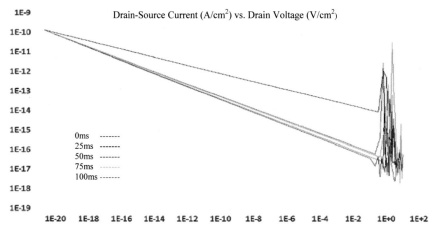

Fig. 8 Graph of I_{DS} versus V_{DS} for time-based study for particle radiation

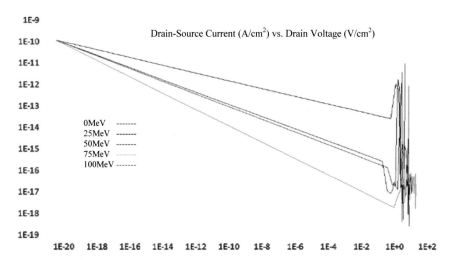

Fig. 9 Graph of I_{DS} versus V_{DS} for dose rate (LET-based study) for particle radiation

5 Conclusions

This chapter presented the study of the radiation effect in the commercial power MOSFET by using the software simulation. For that purpose two types radiation are used which are; gamma ray radiation represent the photon radiation and single heavy ion radiation as to represent the particle radiation. From the simulation result, gamma ray and single heavy ion have given a great impact to the commercial device. Both radiations have caused physically damage to the device structure because the electrical field is changing once the device exposed to the radiation. If the voltage gate applied to the gate electrode increased, the electrical field stress will increase more and in the worst case the device will burnout. For the electrical performance, both devices also have fluctuated the electrical characteristic of the device. From the simulation result, during off state region, the value of drain current is decreasing drastically depends on the dose time and the dose rate.

Acknowledgements The authors are thankful to the Fundamental Research Grant Scheme (FRGS 2016) with the title "A Fundamental Investigation of Radiation Effect on Metal Oxide Semiconductor Field Effect Transistor on Harsh Environment Application".

References

1. Critchilow DL (1999) MOSFET scaling-the driver of VLSI technology. Proc IEEE 87(4):659–667
2. Ning TH (2001) History and future perspective of the modern silicon bipolar transistor. IEEE Trans Electron Dev 48(11)
3. Kaur R, Gupta P, Singh J (2017) A review on power MOSFET device structures. Int J Res Appl Sci Eng Technol 5:208–218
4. Patel MR (2004) Spacecraft power systems. CRC Press, Boca Raton
5. Tang Z et al (2012) A novel structure for improving the SEGR of a VDMOS. J Semicond 33(4)
6. Tang Z et al (2014) A novel terminal structure for total dose irradiation hardened of a P-VDMOS. J Semicond 35(5)
7. Abubakkar SFO, Zabah NF, Abdullah Y, Fauzi DA, Muridan N, Hasbullah NF (2017) Effects of electron radiation on commercial power MOSFET with buck converter application. Nucl Sci Tech 28(3):1–5
8. Reitz Guenther (2008) Characteristic of the radiation field in low Earth orbit and in deep space. Zeitschrift für Medizinische Physik 18(4):233–243
9. Baumstark-Khan C, Facius R (2002) Life under conditions of ionizing radiation. In: Astrobiology. Springer, Berlin, pp 261–284
10. Barth JL, Dyer C, Stassinopoulos E (2003) Space, atmospheric, and terrestrial radiation environments. IEEE Trans Nucl Sci 50:466–482
11. Bourdarie S, Xapsos M (2008) The near-earth space radiation environment. IEEE Trans Nucl Sci 55:1810–1832
12. Yahya EA, Kannan R, Baharudin Z, Krishnamurthy S (2017) An overview of instantaneous radiation effect on MOSFETs for harsh environment applications. In: 2017 IEEE 3rd international symposium in robotics and manufacturing automation (ROMA) (pp. 1–6). IEEE
13. Johnston A (2010) Space radiation effects and reliability considerations for micro-and optoelectronic devices. IEEE Trans Device Mater Reliab 10(4):449–459
14. Schwank JR, Shaneyfelt MR, Fleetwood DM, Felix JA, Dodd PE, Paillet P et al (2008) Radiation effects in MOS oxides. IEEE Trans Nucl Sci 55:1833–1853
15. Stassinopoulos EG, Raymond JP (1988) The space radiation environment for electronics. Proc IEEE 76(11):1423–1442
16. Peretti G, Demarco G, Romero E, Tais C (2015) 3D thermal and mechanical analysis of a single event burnout. IEEE Trans Nucl Sci 62:1879–1887
17. Davidović V, Danković D, Ilić A, Manić I, Golubović S, Djorić-Veljković S et al (2016) NBTI and irradiation effects in p-channel power VDMOS transistors. IEEE Trans Nucl Sci 63:1268–1275
18. Liu S et al (2011) Effects of ion species on SEB failure voltage of power DMOSFET. IEEE Trans Nucl Sci 58(6 PART 1):2991–2997

19. Ferlet-Cavrois V, Binois C, Carvalho A, Ikeda N, Inoue M, Eisener B et al (2012) Statistical analysis of heavy-ion induced gate rupture in power MOSFETs—methodology for radiation hardness assurance. IEEE Trans Nucl Sci 59:2920–2929
20. Privat A, Touboul AD, Michez A, Bourdarie S, Vaille J, Wrobel F et al (2014) On the use of post-irradiation-gate-stress results to refine sensitive operating area determination. IEEE Trans Nucl Sci 61:2930–2935
21. Guide (2007) Mesh generation tools user. Mountain view, California: Synopsys, Inc.; Sentaurus Structure Editor User Guide, Version A-2007.12
22. Version SD (2013) H-2013.03. Synopsys, Mountain View, CA, USA. 2013 Mar

Time Series Models of High Frequency Solar Radiation Data

Mohd Tahir Ismail and Samsul Ariffin Abdul Karim

In this era of Internet of Thing (IoT) and Big Data, observations are collected not as daily, weekly, monthly, quarterly or yearly, but are taken at a finer time scale. These observations are in hourly, minutely, second and then divide into fractions of second have become available mainly due to the advancement in data acquisition and processing techniques. In this chapter, high frequency time series data of solar radiation of every 30 s will be gathered and model using Box and Jenkins methodology. This methodology comprises of model identification, model estimation, model verification and finally model adequacy. Through this Box and Jenkins modelling procedures, the best time series models for high frequency solar radiation data will be suggested for forecasting purposes.

1 Introduction

Recently with the development of high-performance computing (HPC) data acquisition and processing have improved tremendously. Data on a finer time scale, such as minutely or secondly can be collected and analysed in order to obtain more up to date information. This high frequency information/data can help researchers making decision on demand faster and more accurate.

M. T. Ismail (✉)
School of Mathematical Sciences, Universiti Sains Malaysia, 11800 USM Minden, Penang, Malaysia
e-mail: m.tahir@usm.my

S. A. A. Karim
Fundamental and Applied Sciences Department and Centre for Smart Grid Energy Research (CSMER), Institute of Autonomous System, Universiti Teknologi PETRONAS, Bandar Seri Iskandar, 32610 Seri Iskandar, Perak Darul Ridzuan, Malaysia

© The Author(s), under exclusive license to Springer Nature Singapore Pte Ltd. 2020
S. A. A. Karim et al. (eds.), *Practical Examples of Energy Optimization Models*,
SpringerBriefs in Energy, https://doi.org/10.1007/978-981-15-2199-7_6

The high frequency data commonly utilised in financial time series where the information about the stock price every second is being monitored by investors. It also attracts the attention of many researchers to focus on empirical study of high frequency data in financial and economic time series such as stock market index [1, 2] and exchange rate [3]. However, in this present study focus will be on high frequency environmental data. The data that will be utilised is solar radiation data. The main objective of the present studies is:

(a) To propose a model for high frequency solar radiation data using Box and Jenkin methodology
(b) To evaluate the forecasting performance of the propose model.

This chapter is organized as follows. Section 2 discussed related literature review. Section 3 presented the data and the Box and Jenkins methodology. Section 4 is dedicated for results and discussion while finally, a chapter summary will be concluded in Sect. 5.

2 The Literature Review of Related Work

In recent years, many countries have focused on renewable energy such as wind and solar in order to overcome the depleting of fossil fuels and the dangerous nuclear fuels. Moreover, more countries are moving to green energies as a result of the effect of non-renewable energies to the environment. Thus, modelling and forecasting the wind or solar energy is important to increase the efficiency and improve their reliability. This will optimize the operation cost and economic feasibility of these resources.

Many researchers around the world have proposed and developed models of solar radiation using data from their region. These utilised models can be classified as a standalone or a hybrid of two or more models. Some of the models used are regression, machine learning models such as artificial neural network (ANN) and time series autoregressive integrated moving average (ARIMA)/seasonal autoregressive integrated moving average (SARIMA). While the frequency of the data employs are daily or monthly. In addition, some of the measurements used to evaluate the forecasting performances are root mean square error (RMSE), mean absolute error (MAE), mean absolute percentage error (MAPE), the coefficient of correlation (r) and mean bias error (MBE).

Among literatures that working with regression models are Guermoui et al. [4]. They used Algeria daily solar radiation data from 2013 to 2015. They proposed a new method called Weighted Gaussian Process Regression (WGPR) and successful in modelling the data. Then, Muzathik et al. [5] used Malaysia monthly daily average global solar radiation data between 2004 and 2007. They compare the performance of ten regression models and found that simple regression models is the best models. Later, Yap and Karri [6], found that linear regression model performance better in predicting monthly solar radiation data collected from Darwin, Australia.

While, Ghimire et al. [7] achieve in showing that ANN a machine learning model provide better prediction for daily global radiation data from the rich cities of Queensland Australia. They compared with other three machine learning models, namely support vector regression (SVR), Gaussian process machine learning (GPML) and genetic programming (GP) models. Next, Ozoegwu [8] reviewed on the solar radiation data modelling in Nigeria. Based on his study, he found that ANN better in one year ahead forecasting and was verified with the used of one-way analysis of variance (ANOVA).

There are also researchers worked with time series model. Among others are Alsharif et al. [9], Adejumo and Suleiman [10], Fortuna et al. [11] and Sun et al. [12]. Sun et al. [12] conducted an empirical investigation on solar radiation series using Autoregressive Moving Average (ARMA) and Generalized Autoregressive Conditional Heteroscedasticity (GARCH), ARMA-GARCH models using two data sets from China. They discovered ARMA-GARCH models can capture the behaviour of solar radiation observations better than ANN. This result is also supported by Adejumo and Suleiman [10] when they applied to Nigeria daily solar radiation observations. Meanwhile, Fortuna et al. [11] studied the time series properties of solar radiation data from ten stations in USA. Recently, Alsharif et al. [9], construct a seasonal/non-seasonal autoregressive integrated moving average (SARIMA/ARIMA) model to predict the daily and monthly solar radiation in Korea. Their results revealed that for daily data ARIMA models have better performance while for the monthly data SARIMA model is the best.

Apart from that, current study recommended a hybrid model in examining the solar radiation data. For examples, Ozoegwu [13] using a hybrid of nonlinear autoregressive (NAR), a nonlinear autoregressive exogenous (NARX) with ANN and Mukaram and Yusof [14] using a hybrid of SARIMA and ANN. Both papers found that the hybrid models perform better as compare to the individual model. Motivated by previous literatures, a new model is proposed in order to model the high frequency solar radiation data. This model is called an autoregressive fractionally integrated moving average (ARFIMA). This model will be discussed in the next chapter.

3   Data and Methodology

This section presents the data and the method used in modelling the solar radiation data.

3.1   Data

The high frequency solar radiation data was gathered from the Malaysian Meteorological Department from a station at Putrajaya. Three sets of data were collected for every 30 s from 9 am until 6 pm on 21 January 2010. The number of observations is

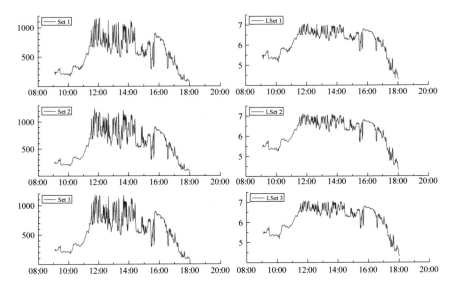

Fig. 1 Time series plot for the global solar radiation data

1076 for each set of data. The time series plots for the three data sets are presented in Fig. 1. The three data sets are transformed into natural logarithm before any analysis being conducted to reduce the variation between the values. However, the trend of the data before and after transformation is preserved. It can also be seen that LSet 1, LSet 2 and LSet 3 are the data after transformation.

3.2 ARFIMA Models and Box and Jenkins Procedure

Long memory is one of the main features of time series data. Long memory also called as long-range dependence basically refer to the level of statistical dependency between two points in the time series. This level of dependency can be seen by using the autocorrelation function (ACF). For a short memory process, the dependence between ACF values at different times rapidly decreases however it is opposite for the long memory process. Thus the further discussion can be found in Palma [15] and Beran et al. [16]. In order to capture long memory, ARIMA model can be extended to ARFIMA model. The general equation for ARFIMA models is given as follows:

$$\phi(L)(1 - L)^d (Y - \mu) = \theta(L)\varepsilon_t \tag{1}$$

where $\phi(L)$ represent the Autoregressive, AR operator p, $\theta(L)$ characterize the moving average, MA operator q, Y is the series and ε_t is white noise. The difference between ARIMA and ARFIMA is the value of d where AFRIMA $-1/2 \le d \le 1/2$

and ARIMA $d \geq 0$. The ARFIMA model's estimation is based on maximum likelihood estimation (MLE) as discussed by Doornik and Ooms [17].

The Box and Jenkins approach [18] for ARIMA modelling will be utilized to find the optimum number of p and q for the ARFIMA model. The approach comprises of four steps. The first step is model identification where the pattern of the ACF will be examined. Then the second step, the model estimation where a few models with difference p and q will be estimated. Next the third step, model diagnostic where the model with all the parameter is significant and fulfil the three assumptions of the residual (normality, no serial correlation, no heteroscedasticity) will be considered for the best models. Finally, the fourth step, model forecasting where the forecasting performance of the chosen model is evaluated based on in sample and out sample forecasting using dynamic and static forecast methods. In dynamic forecasting, previously forecasted values of the lagged dependent series are used in forming forecasts of the current value. While, static forecast calculates a sequence of one-step ahead forecasts, using the actual series. The forecasting performance is evaluated using the root mean square error (RMSE), mean absolute error (MAE), mean absolute percentage error (MAPE) and Theil value.

4 Results and Discussion

In this section, the distribution of the three data sets are being investigated. Then the Box and Jenkins procedure will be applied to find the best model for the high frequency solar radiation data. Based on Fig. 2, all the data sets are skewed to the left with peak near to bell shape. This is justified by the skewness and kurtosis values in Table 1. Overall the data sets did not follow a standard normal distribution.

Next the analyses proceed to the first step of Box and Jenkins procedures that is Model Identification. Based on the autocorrelation (ACF) plot in Fig. 3, the ACF showed persistent pattern of moderately high spikes values and decays at a hyperbolic rate. This indicates the three data sets reveal long memory behaviour. Thus, the ARFIMA model will be utilized to capture long memory feature.

The second step in Box and Jenkins procedure is the Model Estimation. In this step a few models with different values of p for the AR process and q for the MA process will be estimated. Only models with all their parameters are significant will be considered for comparison purposes. The comparison is based on the values of Akaike Information Criterion (AIC), Bayesian Information Criterion (BIC) and Hannan-Quinn Information Criterion (HQC). The model with the smallest values of AIC, BIC and HQC is chosen to the next step. It can be seen from Table 2 that for the three data sets, the best model is the ARFIMA (2, d, 0). This is based on the lowest values of AIC, SIC and HQC. The equation of the ARFIMA model for each data set is given as follows:

$$\left(1 - 1.33L + 0.33L^2\right)(1 - L)^{0.19}(LSet1 - 5.67) = \varepsilon_t \qquad (2)$$

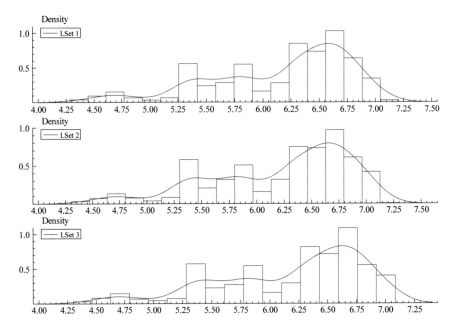

Fig. 2 Distribution of the data

Table 1 Descriptive statistics

Data	Mean	Std. dev.	Skewness	Kurtosis	Observations
Set 1	6.15	0.61	−0.85	2.96	1076
Set 2	6.21	0.63	−0.79	2.80	1076
Set 3	6.18	0.62	−0.82	2.88	1076

$$\left(1 - 1.35L + 0.35L^2\right)(1 - L)^{0.21}(LSet2 - 5.72) = \varepsilon_t \tag{3}$$

$$\left(1 - 1.43L + 0.43L^2\right)(1 - L)^{0.22}(LSet3 - 5.71) = \varepsilon_t \tag{4}$$

The third step in Box and Jenkins procedure is the Model Diagnostic. Looking at Table 3, all the series denoted the residuals have no serial correlation where the Durbin-Watson equal or approach to 2. However, the normality assumption cannot be met. While only data Lset3 showing their residuals have heteroscedasticity. Nevertheless, Fig. 4 reveals the ARFIMA (2, d, 0) for each data set are stable as the inverse root of the AR parameter are within the unit circle.

The final step of Box and Jenkins procedure is the Model Forecasting evaluation. The forecasting performance based on in sample (using all the observation) forecasting is shown in Table 4 and Fig. 5. It appears that the forecasting performance of the static method is better than the dynamic method where the values of RMSE, MAE,

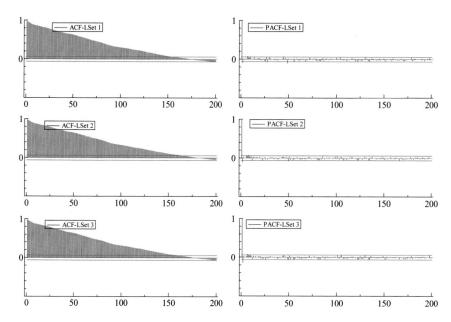

Fig. 3 ACF and PACF plot

Table 2 ARFIMA model estimation

LSet 1

ARFIMA	(1, d, 0)	**(2, d, 0)**	(1, d, 1)	(1, d, 2)
AIC	−2.188	**−2.212**	−2.206	−2.209
SIC	−2.170	**−2.188**	−2.183	−2.181
HQC	−2.181	**−2.203**	−2.197	−2.198

LSet 2

ARFIMA	(1, d, 0)	**(2, d, 0)**	(1, d, 1)
AIC	−2.096	**−2.125**	−2.123
SIC	−2.077	**−2.101**	−2.100
HQC	−2.089	**−2.116**	−2.114

LSet 3

ARFIMA	(1, d, 0)	**(2, d, 0)**	(3, d, 0)	(1, d, 1)	(1, d, 3)	(2, d, 1)
AIC	−2.286	**−2.312**	−2.312	−2.311	−2.312	−2.3127
SIC	−2.267	**−2.289**	−2.284	−2.288	−2.280	−2.284
HQC	−2.279	**−2.303**	−2.301	−2.302	−2.300	−2.302

Table 3 The results of specification tests

Data set	Test	H₀	Statistics
LSet 1	Residual serial correlation	No serial correlation	2.00
	Residual heteroskedasticity	Homoskedastic	1.81
	Residual normality	Multivariate normal	17,082***
LSet 2	Residual serial correlation	No serial correlation	1.988381
	Residual heteroskedasticity	Homoskedastic	4.39
	Residual normality	Multivariate normal	15,576***
LSet 3	Residual serial correlation	No serial correlation	1.96
	Residual heteroskedasticity	Homoskedastic	38.08***
	Residual normality	Multivariate normal	3902***

Note ***Denotes significance at 1% level

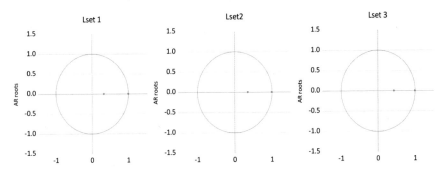

Fig. 4 Inverse roots of the AR parameters

Table 4 In sample forecasting performance

Data set	Measurement	Dynamic	Static
LSet 1	RMSE	0.81	0.07
	MAE	0.71	0.04
	MAPE	11.42	0.72
	Theil	0.06	0.006
LSet 2	RMSE	0.89	0.08
	MAE	0.78	0.04
	MAPE	12.24	0.73
	Theil	0.07	0.006
LSet 3	RMSE	0.87	0.07
	MAE	0.76	0.04
	MAPE	12.05	0.71
	Theil	0.07	0.006

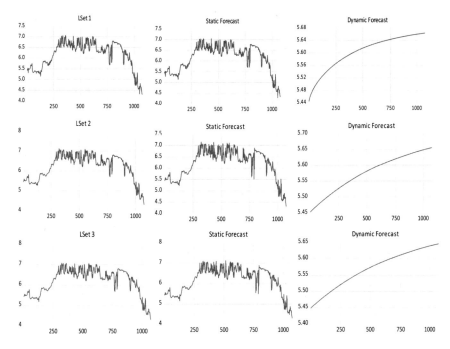

Fig. 5 In sample forecasting fitting

MAPE and Theil is the smallest. This is true for all three data sets. Furthermore, Fig. 5 displayed static forecast follow the similar trend of the observed series. However, this is not true for dynamic forecast where the fitted values revealed upward trend.

Whereas, for the out sample forecasting, 1000 observations are used for modelling and the remaining 76 observations are reserved for forecasting comparison purposes. Again a comparable results as the in sample forecasting was obtained. It can be seen that the forecasting performance of static method is superior to the dynamic method. This is shown by the lowest values of RMSE, MAE, MAPE and Theil values in Table 5 for the three data sets. Moreover, Fig. 6 indicate that 76 observations forecasted by the static method exhibited parallel trend as the observed data. However, 76 dynamic forecasted values demonstrated upward trend.

5 Conclusion

In this chapter, three data sets of the high frequency solar radiation data are taken from the station at Putrajaya, Malaysia. Based on the ACF, it seems that there is a long range dependency in the three data sets. Thus, the ARFIMA model was proposed to model the high frequency data using Box and Jenkins procedure. It was found that the

Table 5 Out sample
forecasting performance

Data set	Measurement	Dynamic	Static
LSet 1	RMSE	0.76	0.10
	MAE	0.69	0.06
	MAPE	14.96	1.37
	Theil	0.07	0.01
LSet 2	RMSE	0.74	0.10
	MAE	0.66	0.07
	MAPE	14.27	1.45
	Theil	0.07	0.01
LSet 3	RMSE	0.77	0.09
	MAE	0.69	0.06
	MAPE	14.95	1.319
	Theil	0.07	0.009

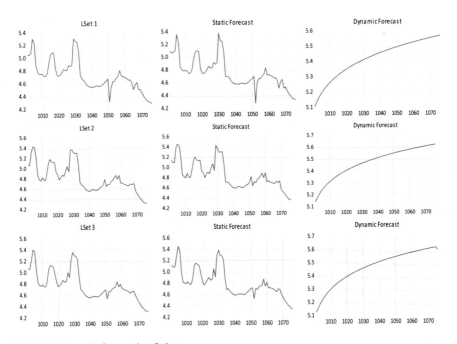

Fig. 6 Out sample forecasting fitting

ARFIMA (2, d, 0) fitted all the three data sets very well. Moreover, the forecasting performance of the static forecast fitted well to the real data sets. Thus, the two objectives of the chapter are fulfilled. It is suggested that for future study, forecasting using static method provides better performance than the dynamic method.

References

1. Sayeed MA, Dungey M, Yao W (2018) High-frequency characterisation of Indian banking stocks. J Emerg Mark Financ 17:S213–S238
2. Zhou X, Pan Z, Hu G, Tang S, Zhao C (2018) Stock market prediction on high-frequency data using generative adversarial nets. Math Probl Eng 2018:1–11
3. Serjam C, Sakurai A (2018) Analyzing predictive performance of linear models on high-frequency currency exchange rates. Vietnam J Comput Sci 5(2):123–132
4. Guermoui M, Melgani F, Danilo C (2018) Multi-step ahead forecasting of daily global and direct solar radiation: a review and case study of Ghardaia region. J Clean Prod 201:716–734
5. Muzathik AM, Nik WBW, Ibrahi MZ, Samo KB, Sopian K, Alghoul MA (2011) Daily global solar radiation estimate based on sunshine hours. Int J Mech Mater Eng 6(1):75–80
6. Yap KW, Karri V (2012) Comparative Study in Predicting the Global Solar Radiation for Darwin, Australia. J Sol Energy Eng 134(3):1–6
7. Ghimire S, Deo RC, Downs NJ, Raj N (2019) Global solar radiation prediction by ANN integrated with European Centre for medium range weather forecast fields in solar rich cities of Queensland Australia. J Clean Prod 216:288–310
8. Ozoegwu CG (2018) The solar energy assessment methods for Nigeria: the current status, the future directions and a neural time series method. Renew Sustain Energy Rev 92:146–159
9. Alsharif MH, Younes MK, Kim J (2019) Time series ARIMA model for prediction of daily and monthly average global solar radiation: the case study of Seoul, South Korea. Symmetry 11:1–17
10. Adejumo AO, Suleiman EA (2017) Application of ARMA-GARCH models on solar radiation for south southern region of Nigeria. J Inform Math Sci 9(2):405–416
11. Fortuna L, Nunnari G, Nunnari S (2016) Nonlinear modeling of solar radiation and wind speed time series. Springer International Publishing, Switzerland
12. Sun H, Yan D, Zhao N, Zhou J (2015) Empirical investigation on modelling solar radiation series with ARMA-GARCH models. Energy Convers Manag 76:385–395
13. Ozoegwu CG (2019) Artificial neural network forecast of monthly mean daily global solar radiation of selected locations based on time series and month number. J Clean Prod 216:1–13
14. Mukaram MZ, Yusof F (2017) Solar radiation forecast using hybrid SARIMA and ANN model: a case study at several locations in Peninsular Malaysia. Malays J Fundam Appl Sci Spec Issue Some Adv Ind Appl Math 2017:346–350
15. Palma W (2007) Long-memory time series: theory and method. Wiley, Hoboken
16. Beran J, Feng Y, Ghosh S, Kulik R (2013) Long-memory processes: probabilistic properties and statistical methods. Springer, Heidelberg
17. Doornik JA, Ooms M (2003) Computational aspects of maximum likelihood estimation of autoregressive fractionally integrated moving average models. Comput Stat Data Anal 42:333–348
18. Box GEP, Jenkins GM, Reinsel GC, Ljung GM (2015) Time series analysis: forecasting and control, 5th edn. Wiley, Hoboken

Index

© The Author(s), under exclusive license to Springer Nature Singapore Pte Ltd. 2020
S. A. A. Karim et al. (eds.), *Practical Examples of Energy Optimization Models*,
SpringerBriefs in Energy, https://doi.org/10.1007/978-981-15-2199-7

Printed in the United States
By Bookmasters